FROM QUANTA TO QUARKS

MORE
ANECDOTAL
HISTORY
OF PHYSICS

FROM QUANTA TO QUARKS

MORE ANECDOTAL HISTORY OF PHYSICS

ANTON Z CAPRI
UNIVERSITY OF ALBERTA, CANADA

World Scientific

NEW JERSEY · LONDON · SINGAPORE · BEIJING · SHANGHAI · HONG KONG · TAIPEI · CHENNAI

Published by

World Scientific Publishing Co. Pte. Ltd.
5 Toh Tuck Link, Singapore 596224
USA office: 27 Warren Street, Suite 401-402, Hackensack, NJ 07601
UK office: 57 Shelton Street, Covent Garden, London WC2H 9HE

British Library Cataloguing-in-Publication Data
A catalogue record for this book is available from the British Library.

FROM QUANTA TO QUARKS
More Anecdotal History of Physics

ISBN-13 978-981-270-916-5
ISBN-10 981-270-916-9
ISBN-13 978-981-270-917-2 (pbk)
ISBN-10 981-270-917-7 (pbk)

Printed in Singapore.

To all my students who made my teaching career so enjoyable. Also to all my colleagues who, not only told me their favourite anecdotes, but also patiently listened to me repeat them back to them.

Preface

When first planning this book I intended it to be part of *Quips, Quotes, and Quanta*, but the material grew too extensive to include in one volume. Also the material in this volume is of a quite different nature. Since I have tried, as much as possible to adhere to a thematic order, there is temporal overlap with *Quips, Quotes, and Quanta*.

In writing this book I have again been helped by numerous colleagues who, knowing of my interest in anecdotes have, over the years, passed many of them on to me. In particular I am most grateful to Professor Werner Israel who not only offered constant encouragement, but also contributed several personal reminiscences. I am also grateful to his wife, Inge Israel for telling me of some humorous incidents involving Werner.

As in *Quips, Quotes, and Quanta*, Professor M. Razavy not only provided me with continuous encouragement, but also with numerous references to historical writings. His encyclopaedic knowledge has been a great help.

Professor Valeri Frolov and Dr. Andrei Zelnikov helped with some biographical material dealing with Soviet and Russian physicists. They saved me considerable time and effort for which I am very grateful.

There are far too many colleagues to list all of those who also contributed in one way or another. I thank them for their continued interest and support.

Over the years I have presented part of this material to various audiences. In particular I have to acknowledge the enthusiastic response of the Science Council of the Alberta Teachers' Association who gave me the opportunity to present almost all of Chapter 1 In November, 2004, at the Banff Centre.

At one stage in the writing of this book I had helpful advice from Joy Gugeler. I thank her for her help.

As in *Quips, Quotes, and Quanta* the translations into English, unless otherwise specified, are my own.

A. Z. Capri
Edmonton, Alberta.
April, 2007.

Contents

Contents

Chapter 1

Prologue

"The scientific theory I like the best is that the rings of Saturn are composed entirely of lost airline baggage." Max Born

Ever since 1958, when I attended a seminar by Arthur Leonard Schawlow (1921 – 1999), I have been on the lookout for humorous and illustrative anecdotes about famous and not-quite-so-famous physicists. At the time I was an undergraduate at the University of Toronto and Schawlow (who shared the 1981 Nobel prize "for his contribution to the development of laser spectroscopy") had returned to his alma mater to talk about and demonstrate the just invented laser. After a fifteen minute period of "pumping" the laser, our eyes were directed to one of the walls in the big lecture theatre of the McLennan Laboratory where, for a fraction of a second, a very bright spot appeared. I was less impressed by this than the 12×2 cm rubies he had grown to make the lasers. I remember very little of Schawlow's talk, but I did retain that maser was an acronym for microwave amplification by stimulated emission of radiation because Schawlow quipped, it should really stand for "money acquisition scheme for expensive research." This was my first exposure to physics humour and started me on my quest for physics anecdotes.

At the time I was not aware that Schawlow was as notorious for his humorous activities as he was famous for his excellent teaching and research. As an example, he and Ted Hänsch built the first "edible laser" out of gelatin and food colour. This dessert led to the later development of a laser that is now used in fiber optics for communication. Art Schawlow also entertained

1

many a class at Stanford with his "Mickey Mouse demonstration". He would place an inflated blue balloon, shaped like Mickey, inside a clear balloon and shoot a laser beam at Mickey. The beam passed through the clear balloon without damaging it and demolished Mickey. This "childish" demonstration was the stimulus for the development of laser surgery to reattach a detached retina in the human eye.

That Schawlow's jokes were always understood as such, even by the people who were the butt, is illustrated by the following. At Stanford he began a seminar with the title "Is spectroscopy dead?" by presenting a long and detailed definition of spectroscopy. In the audience was Felix Bloch who shared the 1952 Nobel prize with Edward Mills Purcell "for their development of new methods for nuclear magnetic precision measurements and discoveries in connection therewith." When Schawlow paused, Felix Bloch challenged him, "Now define dead." After a pause Schawlow responded, "Dead is when chemists take over the subject." Rather than be offended, the chemists in the audience laughed.

In 1970, while still a fresh assistant professor in the physics department at the University of Alberta I participated in a symposium for the under-graduates to familiarize them with current research. Each professor talked for roughly fifteen minutes about his specialty. In my case I talked about Axiomatic Quantum Field Theory. My emphasis was on the idea that in this subject we start with what we believe to be hard and fast facts and then proceed by rigorous mathematics so as to make sure that "even if we can not be sure of the physics, we can be sure of the mathematics". During the question period a student asked, "What if some of your hard and fast facts, your axioms, should turn out to be wrong, what would you do?" I was taken aback. These axioms were something I believed to be beyond doubt. My senior colleague, Professor Werner Israel, came to my rescue with the correct response, "We would be very happy and excited because then we would have some really worthwhile puzzle on which to concentrate."

This little episode highlights two points. The first is that physicists love to solve puzzles that explain how nature works. The second point is more subtle. Progress is more likely to come from an incorrect theory than from no theory at all. This was stated in his typically unembellished manner by Richard Feynman at the *1973 Hawaii Conference on the Development of QED* — Quantum Electrodynamics. "When everything agrees in experiment and theory you are learning nothing. Only when things are crazy and nothing seems to work, then you are learning something."

After I started to teach physics, I found that my students also enjoyed hearing some of the anecdotes about the people whose ideas they had to study. As a consequence I livened up my lectures on quantum mechanics with stories about Bohr, Heisenberg, Schrödinger, Dirac, and Born. In a similar manner, Einstein's numerous foibles and memorable utterances helped to make relativity lectures less austere.

I believe that making such personalities less remote, showing their idiosyncrasies, and making them more like the rest of us answers a need that arises not, as in the case of politicians, from the fact that they wield power, but rather from the fact that their abilities are often so much more developed than our own. Consider the effect on us of the story of Sir Isaac Newton (1642 – 1727) and the Brachistochrone problem.

After Jacques Bernoulli (1654 – 1705) had solved the famous Brachistochrone problem he challenged the natural philosophers and mathematicians of Europe to find the solution within a year. This problem required the mathematicians to find a curve that connects two points, separated both horizontally and vertically, such that if a bead slides under the action of gravity and without friction from the higher to the lower point, the time taken is a minimum. It is a problem in what is today known as the calculus of variation. The mathematical machinery to handle this had just been invented by Bernoulli when he solved the problem and he thought that no one else could solve it. Now Newton, because then master of the mint, had not been active in science for some time and did not hear of the problem until the year was almost up. However, upon addressing the problem he solved it in an extremely elegant fashion and in a very short time, using purely geometric means. Today after three-hundred years of mathematical progress no simpler solution exists. The modern solutions, using the calculus of variation, require several pages to write out. Newton's solution required only one page and was purely geometrical. When Bernoulli was presented with the unsigned solution he had no doubt as to who had created that marvellous piece of work. "I recognize the lion by his paw," was his comment. Incidentally, the solution is a portion of a curve called a "cycloid".

In *Quips, Quotes, and Quanta* I dealt with the physics of roughly the first third of the twentieth century. This might have been called the "European period" of physics. In this book I deal with developments from roughly the 1930s until the mid 1980s and the present.

It is often said that physics is a young person's game. Thus, it is somewhat surprising that many of the names that appeared in *Quips, Quotes, and Quanta* again come up in this book. Good advice to a young physicist is that, "If older physicists tell you that something is possible, believe them. They have tried this and succeded. If they tell you it is impossible, they have tried and failed. In this case you may doubt them." Although original physics ideas require the properties to be found in the young: stamina, courage, imaginination, and lack of knowledge of the impossible, many of the great physicists continued to produce original ideas at the forefront of their discipline even after middle age. Still, young minds seem to have often led the way.

Until about the 1930s European physics dominated. True there were a few Americans such as Michelson, Compton, and Millikan who made seminal and crucial discoveries, but by and large the great theories of the first quarter of the twentieth century were developed by Europeans. Starting in the late 1930s this changed and by the end of the 1940s the USA had emerged as the dominant power in physics. Not only were most of the crucial experiments performed in the USA, but the new crucial theories also were conceived in the USA. True, some of the "American" physicists — having fled from Hitler — were of European extraction, but many of the truly great were born and educated in the USA.

Most of the founders, of what is still called "Modern Physics", are now dead. My generation was the last to mingle with some of them and to record some of these memories before this generation also dies out is timely. I have not any pretensions to having produced a "scholarly" work. Some of the stories recorded here are Gossip, or physics folklore. I have tried to verify as many of the stories as I could, but I have not hesitated to include all stories that I heard and noted down on scraps of paper. Sometimes I had to rely on memory since it would have been rude to take out a pencil and paper and immediately write down what the speaker said.

Chapter 2

The Birthcry of Atoms

"Cosmic rays have for years been regarded as a means of justifying travel to remote areas of the world." H.V. Neher, July, 1955.

The story of cosmic ray physics begins in 1912 with a balloon flight.

For some time physicists had noticed that electroscopes, left sitting, discharged spontaneously. They already knew that an electroscope irradiated with Röntgen's X-rays discharged much more quickly. So, it was only natural that, after careful examination, they reached the conclusion that there was a type of ionizing radiation occurring all the time. The question now was whether this radiation, causing the spontaneous discharge, came from within the earth or from space. One of the earliest attempts to settle this question involved measuring the rate at which an electroscope discharged at the bottom and at the top of the Eiffel tower. The result was inconclusive, but seemed to indicate that the discharge was more rapid near the ground and hence that the radiation emanated from within the earth.

On August 7, 1912 the Austrian physicist, Victor Franz Hess, (1883 – 1964) together with two companions, ascended in a balloon from the village of Aussig, Austria and drifted north to within 50 km of Berlin. The flight lasted six hours and the balloon reached an altitude of almost 5, 000 m. This historic flight was the beginning of the study of what came to be known as "cosmic rays".

Hess decided to measure the discharge of electroscopes more precisely. Between 1911 and 1912 he made ten balloon flights to check this. His historic ascent of August 7, 1912 provided conclusive proof. During the flight Hess recorded the readings of three electroscopes that he had taken along. As the balloon rose, his electroscopes initially discharged more slowly, showing that there definitely was some radiation from the earth. However, above about 600 meter the electroscopes started to discharge more rapidly and near 5,000 meter the electroscopes discharged almost four times as rapidly as at ground level. His published conclusions were, "The results of my observations are best explained by the assumption that a radiation of very great penetrating power enters our atmosphere from above."

Victor Hess was born in Waldstein Castle near Peggau in Austria where his father, Vinzens Hess served as forester to Prince Öttingen-Wallerstein. His mother was Serafine Edle von Grossbauer-Waldstätt. Hess received all of his education, gymnasium and university, in the historic city of Graz. He earned his doctor's degree in 1910 and then moved to Vienna where he worked as assistant to Professor Stephan Meyer. However, it was Egon von Schweidler (1873 – 1948) who familiarized Hess with the most recent developments in the study of radioactivity. Until 1918 when Austria became a republic and abolished all nobility, von Schweidler was Sir Egon von Schweidler. During World War One, Hess used his knowledge of radiation in a practical way; he was in charge of the X-ray division of a reserve field hospital. In 1919 he was appointed extraordinary professor at Graz University. Soon thereafter he visited the USA for two years where part of his research was in the medical application of radium. In 1931 he got the chair in physics at Innsbruck University. His cosmic ray work was recognized in 1936 with the Nobel prize in physics "for his discovery of cosmic radiation". He shared the prize with Carl David Anderson "for his discovery of the positron".

Hess was a staunch Catholic and as such professed his rejection of National Socialism. Thus, after the *Anschluss*, the annexation of Austria by Germany, he was first suspended from his position and shortly after in September, 1938 fired. The Nazis then further forced him to exchange into German bonds all of his Nobel prize money which he had invested in Sweden. This was too much and that very year Hess left for the USA for a position at Fordham University. In 1944 Hess became a naturalized American citizen.

Hess' historic balloon flight started a time of great discoveries and high adventure. Within a short time physicists were flying balloons to ever greater heights and descending to greater depths to test all sorts of ideas about this radiation. One of them, Heinrich Julius Kohlhörster (1887 – 1946) even reached a height of 8,500 m where he found the most intense radiation yet. A Russian physicist, presenting a lecture in French about cosmic, rays came up with this unintended, but appropriate, mistranslation. "I have measured cosmic radiation in the sea and in high mountain ranges; I have measured it on the bottom of lakes and in the high atmosphere, in rock-salt and coal mines and in deepest caverns. Finally I have measured it in hell." He meant to say "in iron" (*dans le fer*) instead of "hell" (*enfer*).

What were these mysterious rays? Most theories at first suggested that they were very high energy photons and as late as 1927 they were still referred to by many German physicists as *Ultragammastrahlen* meaning ultra gamma radiation.

The man who named cosmic rays, Robert Andrews Millikan (1868 – 1953), was a dominant figure in American physics. He was justly famous for his very accurate measurement of the charge on the electron by the so-called "Millikan oil drop experiment". Millikan had also verified Einstein's formula for the photoelectric effect and used this experiment to get an independent measurement of Planck's constant. His reputation as a very careful and resourceful experimentalist was thus well established. These efforts were recognized by the physics world when he received the 1923 Nobel Prize in physics "for his work on the elementary charge of electricity and on the photoelectric effect".

When it came to cosmic rays, Millikan's imagination carried him a bit too far. Misled by a series of numerical coincidences, Millikan thought these were photons that were emitted when one of the following processes occurred. Four protons fused to form a Helium atom, or fourteen protons fused to make a Nitrogen atom, and finally sixteen protons fused to form an Oxygen atom. This caused Millikan, who was an excellent PR man, to proclaim that these photons were "the birth cry of atoms" that were being continuously created, thus proving that the Creator was "still on the job". Later, he used the more imaginative name "cosmic rays". This name has stuck, rather than the more prosaic German name. So, it is no surprise that when a sign with the words "Jesus saves" was put up on the campus of Caltech some students added "and Millikan takes the credit". Another story has it that the unit of PR ability is the "kan", but this unit is far too large, hence the "Millikan".

Millikan was also very frugal. When Jesse DuMond (1892 – 1976) came to Pasadena on his own, to work in the labs of Caltech, Robert Millikan as the head allowed him to have some space, but absolutely refused to have him paid any salary. Many years later DuMond became a full professor and was allowed some assistants.

Also, some time after Walter M. Elsasser (1904 – 1991) had gotten married, Millikan found out and complained to DuMond that a young man without a permanent position had no business getting married. DuMond's response was that it did not make sense to ask a thirty-three year old man to postpone marriage because he did not have tenure since most men in the world were in the same position.

Millikan also had definite ideas about the function of a university and the funding of research. In 1919 he stated, "The creation of research men may not be the function of all universities, but it should certainly be the function of some of them." In 1945 he explained about funding. "My own philosophy has been that since the main function of the federal government is defense, it is clearly correct to take federal funds for Army and Navy research which can immediately be classified as an essential part of our national defense."

After a while Millikan's photon theory became less tenable and there were several indications that cosmic rays were particles. J. Robert Oppenheimer (1904 – 1967) came up with the idea that cosmic rays were neutral particles. Millikan was very opposed. The hostility of Millikan is evident from a letter that Oppenheimer wrote to Ernest Lawrence after Millikan's strong reaction to his 'heretical' suggestion that cosmic ray primaries were neutral particles and not gamma rays. "It was like you Ernest, and very sweet, that you should whisper to me so comforting words about the Wednesday meeting. I was pretty much in need of them, feeling ashamed of my report, and distressed rather by Millikan's hostility and lack of scruple."

When Oppenheimer made his suggestion, he did not know that the particles were charged. This suggestion came from other quarters. But if the particles constituting the cosmic rays were indeed charged, then they should experience an effect due to the earth's magnetic field which would deflect them to one of the two magnetic poles (depending on whether the charge was positive or negative). So, one should expect an increase in the intensity of cosmic rays closer to the poles and away from the equator. This so-called latitude effect would thus definitely establish that cosmic rays were charged particles.

Arthur. H. Compton (1892 – 1962) — winner of the 1927 Nobel Prize in physics "for his discovery of the effect named after him" — set out to prove this once and for all. He enlisted some sixty collaborators from around the world. In 1932 he travelled more than 80,000 km to look for this effect. He visited five continents and crossed the equator several times from as far south as Dunedin, New Zealand to the northern Arctic.

Arthur Compton began his studies of cosmic rays in 1921 with a very simple experiment. He wanted to test Madame Curie's idea that cosmic rays were captured by radium. Accordingly he measured the cosmic ray activity of radium and cosmic rays at the top of the Grand canyon. Then he loaded his equipment on a mule, descended to the bottom of the canyon and repeated the measurements. He found that the activity of cosmic rays decreased, but the activity of the radium remained the same.

Compton began his serious studies of cosmic rays on a visit to India in 1926 to give lectures at the University of Punjab in Lahore. On landing in Calcutta he found out that he was expected to lead an expedition to Kashmir to study cosmic rays. However, no equipment was provided. With the help of C. V. Raman, who was then professor at Calcutta, he managed to construct an electroscope out of the bowl of a hookah and parts fabricated by Raman's mechanic.

Part of the trip into the Himalayas was on horseback and Compton's wife, Betty turned out to be not only an excellent rider, but also a most able assistant. This so inspired Nazir Ahmad, a young professor from the Islamic College, Lahore that he decided to get Muslim girls educated. In 1939 when Compton again visited he found that Nazir Ahmad had indeed kept to his intention and made some progress getting Muslim girls educated.

Compton's extensive travels in search of the latitude effect were successful. Finally in September 1927 he was able to announce that the latitude effect was real, cosmic rays were charged particles and Millikan was wrong.

Millikan did not respond. Instead he set out to check all this himself. As a consequence of the search for this latitude effect, Millikan wound up with a scar above the bridge of his nose.

In order to test for the latitude effect some rather robust quartz electroscopes were required. They had to be able to withstand jostling and shaking without losing calibration. After his young postdoctoral fellow, H. Victor Neher (1904 –) had produced a prototype, Millikan arrived late in the evening directly from a party. He was still wearing his tuxedo, and was anxious to test this apparatus. The electroscope was mounted on a brass cylinder and the readings were taken through a small telescope. The quartz fiber was illuminated by a flashlight. To test the robustness they used Millikan's 1928 Chevrolet. While Millikan drove, Neher made the observations. After a while Millikan wanted to observe himself. So Neher, who was unfamiliar with the car, had to drive. As Millikan was looking through the eyepiece, Neher let out the clutch too fast and the car jerked forward so that the eyepiece struck

Millikan. He was not seriously injured but carried a scar for the rest of his life.

With these electroscopes Millikan set off to the Arctic and sent Neher to South America. Neither one of them detected a latitude effect. At the December meeting of the Association for the Advancement of Science a symposium on cosmic rays was scheduled with the two Nobel laureates, Compton and Millikan invited to participate. As one cosmic ray physicist predicted, it was to be "hot at the Compton-Millikan debate because Millikan has a chip on his shoulder and Compton is ready to knock it off." The debate was indeed fierce with Millikan attacking Compton with everything at his disposal. Compton on the other hand simply presented his data as objectively as possible. After the session the *New York Times* science reporter, William L. Laurence, tried to get the two Nobel laureates to shake hands. Millikan refused.

Now Millikan received even more bad news. Neher had failed to find a latitude effect on the way south, when passing through the region where it should have been greatest, because his electroscope had failed to work. Now, on the way back, he had indeed found the effect. When Neher disembarked in Atlantic City, Richard Tolman (1881 – 1948) from Caltech met him. When asked if he had found a latitude effect, he confirmed that he had. Tolman only tisk, tisked and shook his head. By early 1933 Millikan conceded that the latitude effect was real.

Actually, the results were not entirely conclusive. After all, there might still be other causes for the latitude effect. Also, in a way, finding the latitude effect was fortuitous since the very high energy cosmic rays are hardly deflected by the earth's magnetic field. It was actually what is now called the "solar wind", protons ejected from the sun, that is mainly responsible for the latitude effect and the Northern Lights, not the cosmic rays.

Rapid progress, in studying cosmic rays, became possible after Hans Geiger (1882 – 1947) and his student Walther Müller (1905 – 1979) invented the Geiger-Müller counter, which nowadays is usually referred to as a Geiger counter. With this instrument it was possible to count individual particles. They placed two such counters one above the other, with a bar of gold more than 4 cm thick between the counters and still found that both counters clicked. This definitely established that "the primary cosmic radiation itself consisted of charged particles rather than photons." Photons could not have penetrated through the gold bar.

The expansion cloud chamber, which had already been invented by Charles Thomson Rees Wilson (1869 – 1959) in 1912, permitted physicists to see the individual tracks of particles as they passed through the chamber and provided another important tool. The density of the tracks produced in the chamber made it possible to estimate the mass and energy of the charged particle. If one furthermore added a magnetic field, to cause the trajectory of the particle to curve, one could also see whether the charge was positive or negative and get a further handle on the mass. This tool allowed Anderson to find and identify the positron in cosmic rays in 1932.

Carl David Anderson (1905 – 1991) described his result in a 1933 publication in the *Physical Review* as follows. "On August 2, 1932, during the course of photographing cosmic-ray tracks in a vertical Wilson chamber (magnetic field of 300,000 Gauss) designed in the summer of 1930 by Professor R. A. Millikan and the writer, the tracks were obtained, which seemed to be interpretable only on the basis of the existence in this case of a particle carrying a positive charge but having a mass of the same order of magnitude as that normally possessed by an electron."

Millikan had told Anderson to look at cosmic rays with a cloud chamber. "You may find interesting results." The results were indeed interesting; Anderson had discovered the positron.

During his Nobel speech Anderson talked about his most recent discovery of a new "heavy electron" — the muon. "These highly penetrating particles, although not free positive and negative electrons, will provide interesting material for future study." It is perhaps as a consequence of this, that no prize was ever awarded for the discovery of this particle. Later, Anderson noted that he "received no reaction at all" when he announced the discovery of these heavy electrons. Interesting though is the fact that these particles had been observed all along in cosmic rays and had been interpreted as electrons. It was only after the discovery of the positron that physicists were willing to "see" new particles.

When these new particles were announced, Isidore I. Rabi exclaimed, "Who ordered that?" As late as 1975, V. W. Hughes and T. Kinoshita wrote as follows in *Muon Physics* Volume 1. "Although spontaneously broken gauge theories provide a promising framework within which masses of particles can be treated more rationally, it is not clear at present whether we can understand the differences of the muon and the electron along this line of reasoning. We may still be very far from the day when we will have a satisfactory answer to the question: 'Why does the muon exist?'"

Later, at a time when Rabi was strongly pushing the construction of high energy accelerators, one of the advocates of cosmic ray studies argued that their approach was a much better return on investment since it could be carried out with relatively simple equipment at the top of a mountain. Rabi's response was, "Your argument is unfair. You left out the cost of the mountain!"

Also, many years later Bruno B. Rossi (1905 – 1993) gave an historical talk. After describing his measurements of the muon lifetime — experiments that were performed on cosmic ray muons with simple equipment built for a few thousand dollars — he went on to say, "In the few decades that have elapsed since those days, the field of elementary particle physics has been taken over by the big accelerators. These machines have provided experimentalists with research tools of a power and sophistication undreamed of just a few years before. All of us old timers have witnessed this extraordinary development with the greatest admiration; yet, if we look deep into our souls, we find a lingering nostalgia for what, in want of a better expression, I may call the age of innocence of experimental particle physics."

Incidentally, Anderson and Neddermeyer first called their new particles "mesotons" (meso for intermediate since they were heavier than electrons, but lighter than protons). However, when Millikan saw this he made them change the name to "mesotron" arguing that it should be like electron or neutron. Although Anderson countered with proton, he acquiesced and sent a cable to *Nature* to change the name to mesotron. Nevertheless, Millikan's campaign for this name eventually failed since, after the discovery of the pi meson (pion), the particle was finally called the muon and belonged to a class of particles dubbed mesons, not mesotrons. Here are a couple of quotes in two letters from Millikan to Robert Brode showing his efforts to retain his preferred name.

"I have no idea who started the use of 'meson'. A couple of years ago I wrote to Bethe, about the only man in this country who was using 'meson', and asked him if he did not think it would be desirable if we got together and tried to get some common usage."

"[I] spoke to Swann about this recently in Philadelphia and he feels very vigorously about it that the use of 'meson' is a very unfortunate one, not only because it violates all historical and etymological properties but is also so close in name to a word that has come in French to be used for a house of ill fame, that he will not tolerate its use at all."

Heisenberg who was probably influenced by his father — a professor of Greek languages — pointed out that the Greek word "mesos" has no "tr" in it. Soon afterwards the name "mesotron" was changed to "meson".

A young Japanese physicist Hideki Yukawa (1907 – 1981) had come up with an ingenious idea for the nuclear force. In order to explain nuclear forces, he postulated the existence of a particle with a mass about 200 times the mass of an electron. At the time he was not interested in cosmic rays, but that is where his ideas were first tested and verified.

At first physicists hoped that Anderson's mesotrons were the Yukawa particles. But, muons passed through matter too easily and did not have the strong nuclear interaction needed to be the Yukawa mesotron. The decisive experiment showing this was performed by three Italian physicists hiding from the German occupation forces in Rome.

During the German occupation of Italy in World War Two, three Italian physicists: Marcello Conversi (1917 – 1988), Ettore Pancini (1915 – 1981), and Oreste Piccioni (1915 –) went into hiding from the German occupation forces so as not to be deported to a forced labour camp. They had started a series of experiments in 1941 on the interaction of mesotrons with matter. They managed to scrounge material for their equipment and performed important experiments in Rome. One of their results was that positive and negative mesotrons when stopped in matter behaved very differently. The positive mesotrons decayed as if no matter were present while the negative ones, when captured by heavy nuclei, produced disintegrations, but when stopped by light nuclei acted as if no matter were present. The Yukawa particles should have interacted stongly with all nuclei and produced disintegrations all the time. This meant that these could not be the Yukawa particles. They werre able to publish this result only after the war in 1946.

A collaboration between Homi Bhabha, Herbert Fröhlich, Walter Heitler, and Nicholas Kemmer then explored different versions of the Yukawa theory. Heitler spoke thus of Fröhlich, "Herbert's strength consisted in his wealth of *anschauliche* ideas, by which he grasped the physics without much reference to the underlying mathematics. Formalism was not his strength. ... Such a gift is today unfortunately rare. Of formalists there are enough." On the other hand of Kemmer he said, "[He] gave a very happy balance to our efforts, for Kemmer had mastered the formalism of quantum field theory much better than we." He also said that, "The vector meson theory was developed in three different places at the same time and in almost identical ways: by Yukawa, Sakata and Taketani and by Bhabha and by ourselves."

It took a while for people to realize that the new highly penetrating particle (the muon) was unstable and of secondary origin. The first one to state this clearly was E. C. G. Stueckelberg (1905 – 1984) who already in 1937 wrote in the *Physical Review* the following remarkable statement. "It seems highly probable that Street and Stevenson and Neddermeyer and Anderson have actually discovered a new elementary particle which has been predicted by theory. This particle is unstable and only can be of secondary origin, its mass being greater than the sum of the masses of the electron plus neutrino." When Stueckelberg stated that the particle had been predicted by theory he was referring to one of his own papers in which he had made a somewhat obscure prediction of a new particle.

The most expensive cosmic ray experiment was proposed in 1995 by Samuel Chao Chung Ting (1936 –) who shared the 1976 Nobel prize in physics with Burton Richter (1931 –) "for their pioneering work in the discovery of a heavy elementary particle of a new kind". Ting made his proposal after the United States cancelled the SSC (Superconducting Supercollider) a giant particle accelerator. His idea was to place a huge superconducting magnetic spectrometer inside the space station. Unfortunately the loss of the space shuttle Columbia in 2003 cancelled all shuttle flights for a while. When the flights resumed, all the payloads until 2010 when the shuttle program is scheduled to end, were committed to other proposals. At present (2007), the $ 1.5 billion Alpha Magnetic Spectrometer seems likely never to leave the earth's surface.

Cosmic ray experiments are no longer conducted solely with "relatively simple equipment at the top of a mountain".

Chapter 3

The Dirac Equation

"This [equation] seemed, and still seems to me, the most beautiful and exciting piece of pure theoretical physics that I have seen in my lifetime — comparable with Maxwell's deduction that the displacement current, and therefore electromagnetism, must exist." Sir Neville Mott.

In 1927 at the Solvay conference Niels Bohr asked Paul Dirac on what he was working. To Dirac's response, "I am trying to get a relativistic theory of the electron" Bohr countered, "But Klein has already solved that problem." Bohr was referring to the so-called Klein-Gordon equation, which Schrödinger had first written down, but rejected because it had not given the correct hydrogen spectrum. As mentioned in *Quips, Quotes, and Quanta*, this equation was discovered and rediscovered by several people. So, Pauli dubbed it "the equation with many fathers". This equation, as we now know, describes mesons rather than electrons.

Dirac disagreed with Bohr regarding the relativistic wave equation since the probability density for the Klein-Gordon equation is not positive and thus cannot yield a true probability — something may have a chance of happening or zero chance (i.e. no chance at all), but can never have less than a zero chance of happening. Dirac persisted in his effort to find an equation that gives only a positive probability density and came up with the now famous Dirac equation. This equation predicts all the details of the hydrogen spectrum as well as describing with unparalleled accuracy the properties of electron, and even predicting the correct electron spin.

After he had worked out his equation, Dirac sent a letter to Max Born in which he added a postscript describing the equation that now bears his name. Born showed the letter to Pascual Jordan who in turn showed the result to Wigner with whom he had been working on a two-component relativistic equation. Jordan made the following comment, "It is too bad that we did not discover that equation, but it is wonderful that someone did."

One of the great successes of the Dirac equation was to automatically yield the factor of two that had been so bothersome for the explanation of the Zeeman effect and been solved by Thomas. This implied that the magnetic moment was $eh/4\pi mc$ for all spin $1/2$ particles, even the proton and neutron. [1]

Otto Stern, had perfected molecular beam techniques and used them to measure space quantization — the fact that the magnetic moment of the electron could only point in two directions — as well as the Maxwell distribution of velocities. He now set about to measure the magnetic moment of the proton. Pauli happened to be visiting and asked him what he was doing.

"I am measuring the magnetic moment of the proton."

"That's a meaningless task. Surely you know that the Dirac theory gives you the answer and the magnetic moment is just $eh/(4\pi mc)$."

"Yes, I know what the answer is going to be, but we'll measure it anyway."

Soon afterwards, at a seminar in Hamburg, in which he discussed the experiment then in progress, Stern asked everyone present to write down the result they expected together with their signature. All the theorists, including Max Born, predicted the value given by the Dirac equation. Two months later Stern returned to Hamburg to present the results of his experiments. He had found that the proton had an "anomalous" magnetic moment, that is a value almost three times as large as expected. He then projected the paper with the original predictions and signatures on a screen.

Although an excellent experimenter, Stern was somewhat clumsy. Furthermore he usually held a cigar in one hand. Consequently all fragile equipment was handled by his assistants. If he saw any equipment starting to fall he did not try to stop it, but simply lifted up both hands in despair. His explanation was, "You do less damage if you let it fall than if you try to catch it."

---

[1]Here e represents the charge on an electron, m the mass of the particle involved, and h Planck's constant. See also Chapter 19 of *Quips, Quotes, and Quanta* for details of the Zeeman effect.

A visitor to Stern's laboratory was amazed by the very large number of controls that an experiment then in progress required. When he commented on this fact. Stern replied, "Yes, it's a pity we lost our prehensile tails some millions of years ago. They would come in handy now ." To this the visitor responded, "That wouldn't help you Stern. You would just make your apparatus even more complicated."

Throughout his life, Dirac was enamoured with mathematical beauty, on which he seemed to model his work. In the May 1963 issue of *Scientific American* he repeated, "It seems that if one is working from the point of view of getting beauty in one's equations, and if one has a really sound insight, one is on a sure line of progress."

Dirac's new equation for the electron not only automatically included the spin, so that it did not have to be added as for the Schrödinger equation, but it also predicted correctly all the results that were known for the hydrogen atom. Later in life Dirac commented, "It was really a surprise to me that the spin should come out this way ... and that a particle with spin of half a quantum is really simpler than a particle with no spin at all."

However, the Dirac equation had, what appeared to be, a very serious flaw. Even for an electron all by itself, with no forces acting on it, half of the solutions had negative energy. In fact, there was complete symmetry between negative and positive energy solutions. At first Dirac thought that "Half of the solutions must be rejected as referring to the charge $+e$ of the electron." Soon he realized that this was not acceptable; the negative energy solutions were necessary. This presented another problem. If the slightest disturbance were to occur to an electron with positive energy, would cascade into a lower (negative) energy state and keep cascading to lower and lower energies so that eventually nothing more could happen. The consequences of such cascades would be that all electrons would wind up in this abyss of negative energy and the whole universe would cease to evolve. Of course such a scenario was unacceptable. He therefore thought that his theory should be only an approximation.

Before the end of 1929, Dirac found a solution to this dilemma. According to the Pauli exclusion principle, no two electrons can occupy exactly the same state. So, Dirac postulated that all the negative energy states are filled. Thus, transitions to such a filled state can not occur. The infinite assembly of filled states was called the "Dirac sea". This outlandish idea not only solved one part of the problem, but had a further dramatic consequence. With enough energy an electron might be knocked out of the Dirac sea into a positive energy state and leave a void or hole behind. This hole would respond to an electric field like an electron with a positive charge because if an electric field pushed the electrons in the sea to the right the hole would, as a consequence, move to the left. The hole would also have the same mass as an electron and so behave like an electron with a positive charge. Dirac was aware of these facts, but hoped that these holes would prove to be protons since no positive electrons were known.

Although in the spring of 1929, in the *Zeitschrift für Physik*, Hermann Weyl had suggested the negative energy electrons might be protons he soon found this not to be the case. In November 1930 in (the revised edition) of his book, *The Theory of Groups and Quantum Mechanics* he wrote, "However attractive this idea may seem at first it is certainly impossible to hold without introducing other profound modifications to square our theory with observed facts. Indeed, according to it the mass of the proton should be the same as the mass of the electron; furthermore, no matter how the action is chosen (so long as it is invariant under right and left), this hypothesis leads to the essential equivalence of positive and negative electricity under all circumstances — even on taking the interaction between matter and radiation rigorously into account."

In the preface to this book Weyl had also stated quite clearly, "At present no solution of the problem seems in sight; I fear that the clouds hanging over this part of the subject will roll together to form a new crisis in quantum physics."

The idea that holes were protons was also ridiculed by several prominent physicists such as Bohr and Pauli. The latter had calculated and found that if the Dirac holes were indeed protons then, in an hydrogen atom, the electrons would jump into them in an exceedingly short time and annihilate them both as well as the surrounding matter in a burst of gamma radiation. Thus, with typical sarcasm, Pauli proposed a "Second Pauli Principle" that excluded all such theories because if such a theory were true than any theorist who wanted to expound it would be destroyed in a burst of gamma radiation before he could announce the theory.

Dirac later stated, "I knew that the mass of the holes was the same as the mass of the electron, but I did not have the courage to state this." Nevertheless, he took heed of all these criticisms and in the *Proceedings of the Royal Society A 133*, 60, 1931 wrote, "A hole, if there were one, would be a new kind of particle, unknown to experimental physics, having the same mass and opposite charge of the electron." A year later, Carl David Anderson found the positive electron, the positron, in cosmic rays.

When Heisenberg first studied the Dirac equation, he mused somewhat wistfully, "Well now, the problem of relativity and quantum theory and the electron has been solved by a young Englishman by the name of Dirac and he is so clever it is not worthwhile to compete in one's work with him." Later as the problems with the negative energy states became acute he was less enthusiastic. In a letter to Pauli he wrote (my translation), "In order not to continuously annoy myself with Dirac, I have done something else for a change." In a later letter he again wrote to Pauli *"Das traurigste Kapitel in der modernen Physik ist nach wie vor die Dirac'sche Theorie."* ("The sorriest chapter in modern physics is and remains the Dirac theory.") He went on to state that all of this electron business had made Jordan depressed.

In 1968 Dirac introduced Heisenberg as follows, "I have the best of reasons for being an admirer of Werner Heisenberg. He and I were young research students at the same time, about the same age, working on the same problem. Heisenberg succeeded where I failed."

In their early days, as famous young physicists, Heisenberg and Dirac had travelled around the world. They sailed from America to the Far East on the Shinyo Maru and then later met again in America. Several stories date from that time.

On the portion of the trip that they made together, they agreed to meet at the hotel at Old Faithful in Yellowstone National Park so that they could see some of the geysers go off. When Dirac showed up he had a detailed timetable of all the geysers that were at all accessible and the times at which they went off. Furthermore he had a table of all the distances between the geysers. Using these data he had worked out a route so that it was possible for him and Heisenberg to see almost all of them go off and not waste a minute.

Heisenberg was always a very active and charming man while Dirac was somewhat shy and taciturn. While at sea, Heisenberg danced almost every dance that was held while Dirac sat by himself. During a break in the dancing Heisenberg returned to the table. Dirac, who had been watching all the activities turned to Heisenberg and asked, "Tell me, why do you dance so much?" The ever gallant Heisenberg replied, "When I see a nice young lady I feel compelled to dance."

After a pause Dirac again asked, "Oh, but how do you know she is nice before you dance with her?"

During their trip the boat docked at Hawaii. Since the boat was due to remain there for a few days they decided to visit the University and offer to give a seminar. At the time the chairman, Paul Kirkpatrick was away on sabbatical leave at Cornell University. The acting chairman, Willard Eller was not involved in physics research, but had been forewarned by Kirkpatrick to expect important visitors. Thus it came about that after arriving at the physics department and identifying themselves as Heisenberg and Dirac, they declaring their intention to give talks. Much to their surprise they were refused. Either Eller had forgotten to expect them or, more likely, he could not believe that these two youngsters were who they claimed. A day later another visitor from the Shinyo Maru also visited the physics department. Eller related to him with glee how the previous day two clowns claiming to be Heisenberg and Dirac had offered to give a seminar, but he had, of course, seen through them.

Dirac always abhorred reporters and being interviewed and sometimes went to extraordinary lengths to avoid them. When the boat carrying Heisenberg and Dirac docked in Japan, reporters swarmed on board to interview these two famous men. Dirac, who was standing beside Heisenberg at the railing turned and stepped back. A reporter asked Heisenberg, "Where's Dirac?" Heisenberg simply shrugged and said nothing. The reporters interviewed Heisenberg and left. Dirac was very proud of thus having outwitted the reporters.

Dirac later recalled that while in Japan they visited this beautiful pagoda. Heisenberg, to Dirac's amazement, climbed to the top of the pagoda and with strong winds blowing about balanced precariously on one foot.

There is a paper by Dirac in the *Journal of Mathematical Physics* 1964, "A remarkable representation of the 3+2 de Sitter Group". When Dirac first gave a talk about this result at the Institute for Advanced Studies in Princeton, Res Jost (1918 – 1990) realized that Dirac was not aware that all the representations of the de Sitter Group were known and classified. At a loss how to tell this fact to Dirac in a gentle fashion, he left a note in his mailbox sketching all these results and lingered around the mailbox to see Dirac's reaction. Dirac found the note, read it and stuffed it into his pocket. Unable to contain his curiosity, Jost asked Dirac, "Did you read my note?"

"Yes."

"Well?"

"You should write it up."

After learning that he was to receive the Nobel Prize, Dirac decided to decline the prize because he certainly did not want all the publicity associated with it. Rutherford, however convinced him to accept the prize by assuring him that he would attract far more publicity by refusing the prize than by accepting it. Years later when Jagdish Mehra asked him how he felt about winning the Nobel Prize, Dirac responded, "It was a great nuisance."

The Dirac equation together with the prediction of the positron was the greatest triumph of physics in the decade 1925 – 1935, a decade of incredible triumphs. Later in life, Dirac referred to the period 1925 – 1933 as the "heroic period". Physics was now primed for a whole series of developments destined to overturn old concepts.

John Archibald Wheeler described Dirac's brilliant intellect with the following statement, "Dirac casts no penumbra."

Chapter 4

Quantum Field Theory

"This is the third of four lectures on a rather difficult subject — the theory of quantum electrodynamics — and since there are obviously more people here tonight than there were before, some of you haven't heard the other two lectures and will find this lecture almost incomprehensible. Those of you who have heard the other two lectures will also find this lecture incomprehensible, but you know that that's all right: as I explained in the first lecture, the way we have to describe Nature is generally incomprehensible to us." Richard Feynman, from a lecture published in QED

The most precise theory mankind has ever known is quantum electrodynamics or QED. Its predictions have been tested experimentally to the incredible accuracy of one part in one hundred billion — that is, to an accuracy of one billionth of a percent. Even so, more than once, physicists were ready to give up on this and other quantum field theories. QED has truly had many ups and downs, only to survive as a most robust and wonderfully accurate theory — the most accurate in all of human history.

The Dirac equation provided the first phase in the development of this theory. The new problem was to consider the interaction of the electromagnetic field with Dirac's electrons. In doing the calculations Dirac soon came up against a serious problem. The interaction changed the mass of the electron by an infinite amount. This ridiculous result was totally unacceptable. To remedy this physicists had to resort to some drastic mathematical manipulations. This was the beginning of what was at first sarcastically called "subtraction physics" but is now called "renormalization theory". Renormalization Theory was completed, independently of each other, by three different individuals: Julian Schwinger (1918 – 1994), Richard Feynman (1918 – 1988), and Sin-Itiro Tomonaga (1906 – 1979).

From 1926 until the early 1930's there was a great increase in interest and enthusiasm for quantum field theory. Then, problems arose. Perfectly reasonable quantities, when calculated to a first approximation gave results in good agreement with experiment. However, when calculated to greater accuracy yielded, instead of small corrections, *infinite* corrections. These totally nonsensical results were disheartening. A majority of physicists even considered these difficulties to render the whole enterprise, not only wrong but even worse, leading in the wrong direction.

The Dirac equation and Dirac's solution of the negative energy states, by introducing the Dirac sea, had created another host of problems. First there was this ugly concept of a sea of infinitely many electrons in states of negative energy. Even after some successes of this approach, Pauli in 1936 at Princeton stated, regarding the Dirac sea. "Success seems to have been on the side of Dirac rather than of logic." Earlier at the Seventh Solvay Conference in 1933, after a talk by Dirac, he had commented. "The theory of holes always seemed very interesting to me, on account of the essential role played in it by the exclusion principle. Whereas the principle was formerly only an isolated rule, of which the validity was independent of those of the other bases of quantum theory, the theory of holes, introduced by Dirac in order to escape the difficulty of negative masses, would have been impossible if we had not wished to exclude all wavefunctions which are not antisymmetric. However, the general aspect of the theory is not satisfactory, owing to the manner in which it is obliged to use the concept of infinity."

Then, as Dirac himself discovered, there was an even more serious problem. If you allowed a Dirac electron to interact with the electromagnetic field that it produced by virtue of being charged, the energy of this "self-interaction" would add to the electron's mass. However, the amount of mass added, by this self-energy, was infinite. Now, this problem had already existed for the classical (pre-quantum) theory of the electron and had been "resolved" by giving the electron a finite size such that all of its mass was due to the electromagnetic self-energy. The same sort of trick would also work in the quantum case. There was only one problem, the theory would then be internally inconsistent. It would no longer satisfy the conditions required by relativity theory. Results like these caused Heisenberg to say later in life, "Up till that time (1928), I had the impression that, in quantum theory we had come back into the harbour, into the port. Dirac's paper threw us out into the sea again."

In 1927, even before his equation, Dirac had published a paper *The quantum theory of the emission and absorption of radiation.* This was the actual start of quantum field theory. In 1929 Heisenberg and Pauli produced the first fully relativistic quantum field theory. Just like Dirac's earlier work, this also produced an infinite mass for the electron.

To remedy the problem of infinities for physical quantities became the problem of the day for theoretical physicists. In 1932 Enrico Fermi wrote a paper, *The quantum theory of radiation,* that explained Dirac's 1927 paper in simple terms. Bethe later wrote, "Many of you, probably like myself, have learned their first field theory from Fermi's wonderful article".

Dirac himself was not happy with his theory but, the fact that his equation, when applied to the electron in the hydrogen atom, gave all the details of the hydrogen spectrum including corrections to the results found with the Schrödinger equation, showed that it was indeed the correct equation for the electron. The problem of the infinite self-energy remained. At the seventh Solvay Conference in 1933 Dirac and Sir Rudolf Peierls (1907 – 1995) presented ideas that would eventually lead to a solution, the so-called "Renormalization Theory". Nevertheless, the solution to this plague of infinities was delayed until the 1940's.

In the mid-forties, Hans Bethe performed what could only be considered a calculation of great courage and even greater physical intuition. He used a nonrelativistic approximation of quantum electrodynamics and chopped off (threw away without justification) the expressions that would have become infinite. He did this by simply making the smallest distance in the calculation the same as the de Broglie wavelength of an electron with a momentum given by its mass times the speed of light. This wavelength is called the Compton wavelength. The quantity Bethe calculated was a frequency of 1040 Megahertz and is called the "Lamb shift". It refers to the shift in the frequency of light that is emitted in an hydrogen atom in a transition from the first excited state to its lowest state. Rumor has it that Bethe did the calculation on a railroad trip from New York to Schenectady where he was going to consult for General Electric on the design of betatrons.

According to Willis E. Lamb Jr.(1913 –) his "first contact" with the problem of the hydrogen fine structure occurred while still in high school. He had met Albert Michelson at a chess tournament in Pasadena. Although still young, Lamb was already quite good at chess. At this time, Michelson was working at Caltech and was very fond of chess. Since Michelson had much earlier found complexity in some of the lines of hydrogen this was Lamb's first contact with this problem, although at the time he was not aware of this.

The physical chemist Gilbert Newton Lewis who, in 1926, invented the name "photon" had advised Willis Lamb to become an experimentalist rather than a theorist because, "a theorist without a good idea is useless, whereas an experimentalist can always go into the laboratory and polish up the brass". He also once told a graduate student, "Damn, if I understood the question, I'd know the answer."

In 1947, Willis E. Lamb and R. C. Retherford measured in a series of experiments, what is now called the Lamb shift, and verified Bethe's calculation. In their paper [1] they state, "The hydrogen atom is the simplest one in existence, and the only one for which essentially exact theoretical calculations can be made on the basis of the fairly well-confirmed Coulomb law of interaction and the Dirac equation for an electron." In this paper they showed that there is a small energy difference for the two levels in hydrogen corresponding to $n = 2, j = 1/2$, but having $l = 0$ or $l = 1$. If one uses the Dirac equation with just one electron, these two levels have exactly the same energy.

In 1950 Arnold Sommerfeld sent a letter to Lamb congratulating him on his measurement and indicating that he (Sommerfeld) was the "eighty-one year old great-grandfather" of the hydrogen fine structure.

The 1955 Nobel prize in physics was shared by Willis Eugene Lamb, "for his measurement of the frequency shift named after him", with Polykarp Kusch (1911 – 1993) "for his precision measurements of the magnetic moment of the electron". Kusch had worked closely with I. I. Rabi and was a professor at Columbia. When he told a student that he had received the Nobel Prize, the student's response was, "Who are you?"

When Oppenheimer, in the 1960s, asked Lamb if he did not regret having missed the Los Alamos experience, the latter said that staying at Columbia was the right decision for him since otherwise he would never have worked on the hydrogen fine-structure experiment. Oppenheimer was less than pleased.

Lamb's experimental verification of Bethe's crude calculation renewed interest in quantum field theory. It led to more refined theoretical attempts and culminated in the next stage of field theory. Physicists now realized that

---

[1] *Physical Review* **72**, 241, (1947)

field theory had to be taken seriously and set out to find a way to control those horrible infinite quantities.

Another important result in QED (quantum electrodynamics) was due to Victor Frederich Weisskopf (1908 – 2002), Pauli's erstwhile assistant. He showed that unlike the nonrelativistic case, the electron self-energy diverged not linearly, but only logarithmically. This was very useful in the development of renormalization theory. Actually Weisskopf had made a sign error (something that is fairly common in any lengthy computation) in his calculation. Thus, two terms added instead of almost canceling. The result was corrected by Wendel H. Furry (1907 –). With a logarithmic divergence the self-energy could be made finite by assuming that the electron had a very tiny radius, very much smaller than the classical electron radius. In fact the radius would be so small as to have no effect on other physical results. Still this was not acceptable because the theory would not be completely relativistic.

Regarding prediction, Weisskopf repeated a statement by H. F. von Ploetz, the Finnish minister of foreign affairs. "It is difficult to make predictions, especially when they concern the future."

As a consequence of computations, like those by Bethe and Weisskopf, the solution to all these difficulties was found: Renormalization Theory. So, how did all this come about?

Julian Schwinger was born in New York City in February of the year in which World War One ended. He progressed rapidly through the school system. At the age of 16, while a student at City College of New York he wrote, but did not publish, a paper on quantum electrodynamics and encountered the infinities that beset this theory. He stated that this term had to be discarded. This was, as stated earlier, the age of "subtraction physics". Isidor I. Rabi persuaded Schwinger to transfer to Columbia University where he completed his thesis for the Ph.D in 1936. However, he did not receive his degree until 1939. Apparently, Schwinger had failed to attend the mathematics lectures and had not enough credits. George Uhlenbeck (1900 – 1988) happened to be at Columbia in 1938 and Rabi told Schwinger that he had to take Uhlenbeck's lectures to get his degree. Unfortunately these lectures were in the morning and Schwinger, being a night owl, did not attend. Eventually Rabi and Uhlenbeck agreed that Schwinger could receive a grade if he passed an exam. Rabi insisted that it be a tough exam. Schwinger not only passed, but as Uhlenbeck said, "Of course, he knew everything." Schwinger got an A.

After his Ph.D. Schwinger moved to the University of California at Berkeley first as a National Research Council Fellow and then as J. Robert Oppenheimer's assistant. In 1941 he moved to Purdue University, but left on a leave of absence in 1943 for the MIT radiation lab to do war research. Schwinger stayed there until 1945. He was morally repelled when he realized the destructive power of what was being created at Los Alamos.

At the radiation lab he worked on radar. Also, in this less restrictive atmosphere he could work at night. Unlike everyone else, he did not restrict his work to any specific project, but rather came in evenings and wandered through the offices to see if anyone had an interesting problem. When he found one, he would solve it and leave the solution on that person's desk. This fact soon became well known. So much so, that Mark Kac (1914 – 1984), who was working at the same lab, received a request from a friend to leave a particularly difficult problem, which he had so far been unable to solve, on Schwinger's desk so that Schwinger might solve it. Kac did as requested and sure enough the next morning he discovered forty-two pages in Schwinger's meticulous handwriting with the solution. So Kac mailed the solution to his friend.

About a week later his friend telephoned to ask Kac to tell Schwinger that the solution was wrong. He could say so because there was a special case of this problem which was particularly simple to solve and Schwinger's solution did not reduce to this simple solution in this special case. With trepidation Kac approached Schwinger and intimated that the solution that Schwinger had presented was wrong. Schwinger simply responded, "No!"

At this point Kac decided to do his own investigation and sat down to check Schwinger's computation. The work was a marvel of accuracy. Schwinger had even listed the references from which he had taken some integrals. Kac went and checked. The book used by Schwinger was an older British table of integrals and the indefinite integrals were not written simply as $\int$ but instead had an inserted upper limit of x like this $\int^x$, but no lower limit. When Kac checked further he found that this was the source of the discrepancy. Schwinger had taken the lower limit of this integral to be zero whereas in fact it was some other totally arbitrary constant. Thus, the mystery was explained.

That Mark Kac had a sense of humour is revealed by the following. Mark Kac was of Polish descent, and used this fact to try and help a student on an oral exam. When the student was asked what kind of singularity a certain function had, Kac pointed vigorously to himself. When the student still did not understand, Kac asked, "What am I?" The student was still at a loss so Kac supplied the answer, "A simple Pole."

The mature Schwinger had as a guiding principle, "Thou shalt not entangle that which is known and reliable with that which is unknown and speculative".

Julian Schwinger always was the master of complicated calculations. He also championed the use of Green's functions. During a seminar by one of Schwinger's students, Green's functions and other complicated mathematical machinery just flew. Fermi, who by then was fifty-five and was in the audience, confided to Segré after the talk, "I am getting old. I understood nothing except the last sentence, 'Finally we arrive at Fermi's Theory of beta decay'." On another occasion, Fermi stated some result and Schwinger asked where this was to be found.

Fermi: "It's in every book on Quantum Mechanics."

Schwinger: "Name one."

Fermi: "It's in Rojanski." It wasn't.

Raymond F. Streater (1936 –) has told the following story about his first meeting with Schwinger in 1963. At the time Streater was in charge of organizing the seminars at Imperial College, London. When he had already lined up speakers for the next few weeks, his former thesis supervisor, Abdus Salam (1926 – 1996) told him that Schwinger would be in London in two weeks and Streater should politely ask one of the speakers to change his talk to a later date so that Streater could invite Schwinger. Everything went fine, except that Schwinger declined explaining that he was in London "to get some rest."

Salam was surprised, "You did not take, 'No' for answer?"

When Streater made a face, Salam agreed that the young man could not argue with Schwinger. He then said that he would handle the matter. Schwinger did agree to give a talk and Streater learned how the matter was handled.

Salam had agreed with Schwinger that he should have a rest in London, but argued that he, Salam, should handle the costs. Accordingly, Schwinger was met at the airport with a fancy limo and taken to a fancy hotel where he was left alone for three days. After that time, Salam phoned to ask how everything was and how Schwinger was enjoying the London shows. Schwinger affirmed that the hotel was fine, but that he had not left the hotel. Salam then informed Schwinger that once in London, you had to take in some of the sights. Schwinger agreed to go to dinner with Salam and P.T.

Mathews the following evening.

To show Schwinger the sights they took him to a private dining club in Soho using the unexpired membership card of a previous visitor. The doorman asked Schwinger and Salam to sign in; Mathews had assumed the name on the card and did not have to sign in. At the time, the law required that one had to sign in if the club had topless waitresses. Schwinger hesitated since he did not relish the idea that it might become known that he had attended such an establishment. Salam laughed, "You don't need to use your own name. Watch" and he signed, "Abdullah." Finally Schwinger also signed. The next day Schwinger agreed to give a seminar. When Salam asked him what name he had signed, Schwinger answered, "P. T. Mathews."

In the spring of 1948 there was a major conference on QED (quantum electrodynamics) in the Pocono Mountains at which, in a series of lectures, Schwinger presented his new results and outlined renormalization theory. Robert Oppenheimer was very enthusiastic and declared, "Now we have a field theory in hand". This brought the following outburst from I. I. Rabi, "What the hell shall I measure now?" Of course, as stated earlier, Rabi had measured the magnetic moment of the electron to high accuracy and thus provided one more clue for electrodynamics.

After this meeting, Schwinger was a somewhat reluctant celebrity. Thus, when he arrived at the "Old Stone on the Hudson" meeting he was immediately rushed upon to report what he had been thinking about. Less than pleased, to be assaulted like that, Schwinger for once replied with uncharacteristic facetiousness, "The Harvard group was not thinking, it was writing".

After Schwinger's presentation, at the Pocono Conference, followed a talk by Feynman. By now, everyone seemed too tired to follow. Bohr gave Feynman a rough time after the talk and Bethe had to console him. That Feynman was more than just brilliant was soon to become evident. This is what Marc Kac said about him years later. "There are two kinds of geniuses, the ordinary and the magicians. Richard Feynman is a magician of the highest caliber."

Also that same spring, Sin-Itiro Tomonaga sent the first two issues of *Progress of Theoretical Physics* to Bethe. In the second issue was a translation of Tomonaga's 1943 article on Quantum Electrodynamics. In this article, he had set the proper foundations for QED, in roughly the same manner as Schwinger. Soon afterwards Oppenheimer received a letter from Tomonaga in which he sketched the further progress that the Japanese physicists had made more recently. Oppenheimer then invited Tomonaga for a visit to Princeton. This was followed by a succession of Tomonaga's students coming to Princeton and Cornell.

Tomonaga's father had been a philosophy professor at Kyoto university. This had influenced Tomonaga's attitude to philosophy. He did not believe one should follow a given philosophy too closely. "Einstein must have much respected Ernst Mach, but he did not obey him blindly. That is really the attitude of the true physicist. It is very dangerous to label each physicist as either Machist or anti-Machist."

In his autobiography, *Disturbing the Universe*, Dyson later wrote, "Amid the ruin of the war, totally isolated from the rest of the world, Tomonaga was, in some respects, ahead of anything existing anywhere else at the time." Tomonaga also said that his postwar isolation was good in one sense. It allowed him to concentrate on his own ideas.

After spending a couple of years at Princeton, Tomonaga returned to Japan. When his ship docked reporters swarmed aboard to interview the famous physicist. Nonplussed, Tomonaga stated that they had the wrong man and pointed to a stranger standing on the upper deck.

Here are two quotes from Tomonaga. "What I am today I owe entirely to sake. I was not blessed with physical and mental strength as a child, but sake cured completely my inferiority complex."

"If you formulate the problem correctly, that is, if you ask the right question, the answer emerges spontaneously."

In Japan, the young theorist, Hideki Yukawa (1907 – 1981), had also attacked these infinity problems. His efforts were without success, but, as mentioned, his classmate Sin-Itiro Tomonaga solved them. Later, in his biography, *Tabibito*, Yukawa described his efforts. "Each day I would destroy the ideas that I had created that day. By the time I crossed the Kamo River on my way home in the evening, I was in a state of desperation. Even the mountains of Kyoto, which usually console me, were melancholy in the evening sun. ... Finally, I gave up that demon hunting and began to think that I should search for an easier problem."

In his search for an "easier problem" Yukawa started to think of the problem of the forces inside a nucleus, especially those between protons and neutrons. As already related, attempts had been made to understand these nuclear forces in terms of the Fermi interaction. Since then two Russian physicists, Igor Tamm (1895 – 1971) and Dmitri Ivanenko(1904 – 1994) had shown in 1934 that Fermi's theory could either explain beta decay or the strong forces inside the nucleus, but not both. In fact there was a discrepancy of a factor of a million million between the two effects. An interesting aside is that later four of the greats of physics left signed statements on the walls of Professor Ivanenko's office in Moscow.

"Physical law should have mathematical beauty", P.A.M.Dirac (1956)

"Nature is simple in its essence", H.Yukawa (1959)

"Contraria non contradictoria sed complementa sunt", N.Bohr (1961)

"Time precedes existence", I.Prigogine (1987)

The paper by Tamm and Ivanenko was a revelation for Yukawa. In his autobiography he later described this. "I was heartened by the negative result and it opened my eyes, so that I thought: let me not look for the particle that belongs to the field of nuclear force among the known particles, including the new neutrino. ... When I began to think in this manner I had almost reached my goal. The crucial point came to me one night in October. ... My new insight was that [the range of the force] and the mass of the particle that I was seeking are inversely related to each other. Why had I not noticed that before? The next morning I tackled the problem of the mass of the new particle and found it to be about two hundred times that of the electron."

This may also be why he commented to his colleague Takehiko Takabayasi, "Great men are lucky, for example Newton and Einstein. I wonder why that is so."

Later in life he pleaded the case for leaving scientists in peace. "It must be admitted that scientific study, in a broader sense, exists for human beings and can always have totally unexpected social consequences. However, if citizens value scientific investigation, I would ask them to leave the scientist in the laboratory, and not to drag him out into the complicated world. A long time ago I ceased to be a no-name Gombei [John Doe]; now no one will leave me alone. I am not unhappy to feel that I am of some worth, but neither can I deny that it is a heavy load for me to bear." The "Gombei" he refers to here is a nickname his other students used since, except for his scholastic abilities, he was so entirely average.

Until the discovery of the pion, the lifetime of the mesotron (muon), observed in cosmic rays, and its interaction with matter were fraught with experimental and theoretical difficulties. Later in life Yukawa recalled this period. "I was full of confidence at the outset when the meson had not yet been discovered; but after it was discovered and its properties began to be understood, curious to say, I gradually lost my self-assurance. It was partly a matter of age: in my twenties I had not doubt at all, but in my thirties I could not help wavering."

By 1939, the thirty-three years old Yukawa was already world famous and everyone interested in cosmic rays was familiar with his work. This is when he started his world trip, which had to be cut short due to outbreak of the war in Europe. At the advice of the Japanese consulate he left Germany and sailed from Hamburg to New York. From there he continued by train to San Francisco where he again boarded the Kamakaru Maru for Yokohama. On his trip across the USA he met many of the leading physicists of this country. From his meeting with Einstein he wrote, "I had the impression that Einstein still believed that the present quantum theory was incomplete and that there should exists some correct continuous theory to replace the present one." In Caltech Yukawa met and dined with Anderson. "Anderson has produced brilliant achievements in physics, but there is no pretentiousness about him."

After the war, Yukawa visited London and met Dirac for the first time. After bowing, in Japanese fashion, and some polite remarks he ventured, "Professor Dirac in your last paper there is a sign error in one of your equations." Dirac then inquired, "But the result is correct?" After receiving an affirmative answer, Dirac continued, "In that case there must be an even number of sign errors."

As a student, and even later in life, Yukawa frequently had doubts. In *Tabibito* he wrote, " ... a little before my graduation from Kyoto University, I began to have misgivings. If I were to continue doing physics, it might come to nothing. I was so pessimistic that I even thought of becoming a priest."

When Hideki Yukawa was already an older man, he gave a public seminar on physics. At the end of the talk someone in the audience asked whether he thought there would ever be an end to physics.

"Yes, of course, all human endeavours must come to an end."

"What kind of an end do you foresee?"

"Possibly the whole world blows itself up, but this is unlikely. It is much more likely that people will just get tired of it."

One of Yukawa's favourite quotes was, "I know that a fish swimming in the river is happy though I have no positive proof of it." This illustrates his attitude to things that might not be observable.

Einstein had a similar quote. "What does a fish know about the water in which he swims all his life?"

Taketani had this to say about the first course Yukawa taught on quantum mechanics. "There were no particular characteristics to Yukawasan's lectures, which followed Dirac's textbook for the most part. His voice was as gentle as a lullaby and he spoke with little emphasis — it was ideal as an invitation to sleep."

Another key figure in the development of modern QED was Freeman Dyson (1923 –). He had started out as a mathematician and even published some seminal work in that subject, but later switched to physics. His explanation for this switch was as follows. In Cambridge, Harish-Chandra, became Dirac's assistant. While on a walk with Dyson and Nicholas Kemmer he declared, "I am leaving physics for mathematics. I find physics messy, unrigorous, elusive." To this Dyson replied, "I am leaving mathematics for physics for the same reasons." They both did as they said and later were reunited at the Princeton Institute for Advanced Studies where Harish-Chandra thrived as a famous mathematician and Dyson as a famous physicist.

Early in his career, after an unfortunate mishap, Dyson abandoned any idea of doing experimental physics. He was trying to set up the equipment for the Millikan oil drop experiment to measure the charge of an electron. In so doing he inadvertently made a bad connection and the high voltage knocked him out. After he revived he had lost all interest in doing further experimental work.

In the summer of 1947 Dyson came from Britain to Cornell in the USA. After he met Feynman he moved with him to Princeton to do the research for his thesis. He never completed this work and never received a Ph.D. since he achieved his greatest work and fame before this. The following summer he had a Commonwealth Fund Fellowship that paid him for a summer vacation during which time he was expected to travel across the continent to see more of it. When the term ended in June, Bethe arranged for him to attend a summer school at Ann Arbor, Michigan where Schwinger would lecture on QED. In the intervening two weeks he travelled with Feynman by car from Cornell to Albuquerque, New Mexico. Their trip was interrupted for a day by a cloud burst that caused a flash flood and washed out the road. Consequently they spent the night, discussing physics, in a motel in the small town of Vinita. After he left Feynman in Albuquerque, Dyson took a bus back to Ann Arbor and attended Schwinger's lectures. However, in usual Schwinger fashion, the lectures were beautifully polished and Dyson found it hard to see everything that was contained in them. On the other hand, Schwinger was also very approachable and Dyson frequently talked to him outside of lectures. Thus, he was able to get a good understanding of what was involved.

Schwinger was frequently accused of presenting work that was "like a cut and polished diamond with all the rough edges removed, brilliant and dazzling." I have myself experienced this phenomenon. During his lecture everything was perfectly clear and logical, but when I went to use my notes from his lectures I had great difficulty in reproducing his results. Another quote about the excessive elegance of Schwinger's lectures is the following. "A marvel of polished elegance, like a difficult violin sonata played by a virtuoso — more technique than music." Schwinger pleaded "not guilty" to these accusations and tried hard to put his beautiful work in context with experiment and historical precedent. I must admit, however, that to me it seemed that some of the things he did could only be done by someone with the genius of a Schwinger.

After the summer school, Dyson travelled to San Francisco and at the beginning of September headed back east with a three-day stopover in Chicago. Although he had not thought of physics for some time now, the ideas of Schwinger and Feynman jelled and he realized that he could show that they were really the same thing. He planned the paper to be "The Radiation Theories of Tomonaga, Schwinger and Feynman". When he came to Chicago to talk about QED à la Schwinger and Feynman, he was careful to extol the virtues of Schwinger's work, but also went on to emphasize that Feynman's methods were more useful and more illuminating. After his talk Teller asked the first question, "What would you think of a man who cried 'There is no God but Allah, and Mohammed is his prophet' and then at once drank down a great tankard of wine?" Dyson was speechless, but Teller immediately answered his own question. "I would consider the man a very sensible fellow."

Back in Princeton Dyson had a really hard time when he presented his results. For, although Schwinger had been a student of Oppenheimer, the latter was not at all receptive to Schwinger's work and was actually hostile to the ideas of Feynman. Along with Bohr, Oppenheimer believed that entirely new physical theories were required. Accordingly, he seemed not to really listen and attacked Dyson's ideas mercilessly.

Fortunately, a short while later, Bethe came to Princeton and during his talks made comments like, "Well. I'm sure that Dyson has explained all this". At which point Dyson would say that he had not yet got this far. Bethe concluded by saying that Feynman's way of calculating was by far the best and people would have to learn it, if they wanted to do serious physics. After this Oppenheimer actually listened carefully to Dyson's lectures and after the fifth lecture left a note in Dyson's mailbox, "Nolo contendere. R.O." (I do not dispute. R. O.)

Although Dyson is best known for his work in quantum electrodynamics, his unorthodox ideas and wit are also well known and illustrated by the following quotations.

"The conspiracy of all infinities to cancel each other must be taken into account."

"The history of physics is littered with the corpses of dead unified theories."

"If you don't have a nasty obituary you probably did not matter."

"Most of the papers which are submitted to the *Physical Review* are rejected not because it is impossible to understand them, but because it is possible. Those which are impossible to understand are usually published."

After Dyson's proselytizing, Feynman's method became the quantum field theorists' gospel. Richard Feynman's computational techniques had made it possible for anyone with very little knowledge of field theory, but enough stamina, to compute any result to arbitrary accuracy. In fact, more mathematically inclined physicists like Res Jost (1918 – 1990) were disturbed by this trend to ignore epistemological questions and simply compute. So much so that he complained, "In the late forties and early fifties under the demoralizing influence of perturbation theory the mathematical sophistication of a physicist was reduced to a rudimentary knowledge of the Greek and Latin alphabets". Even Julian Schwinger, who along with Feynman and Tomonaga created modern quantum electrodynamics, had this to say regarding Feynman's very efficient way of computing with the so-called Feynman diagrams, "Like the silicon chip of more recent years, the Feynman diagram was bringing computation to the masses."

The power of Feynman's method is also well illustrated by a story that he told in one of his talks. At the January meeting of the APS (American Physical Society) in 1949, Murray Slotnick presented a talk in which he compared two different kinds of electron neutron interaction. His computations were extremely complicated and in the end he showed that they gave different results for scattering in the forward direction. To Slotnick's chagrin, Robert Oppenheimer stood up and stated that the result must be wrong since it disagreed with Case's Theorem. When Slotnick admitted that he had never heard of the theorem, Oppenheimer told him that he could learn about it by attending Kenneth Case's talk the next day. Feynman then spent the night working out Slotnick's result. In his talk (*Physics Today* page 48, September

1996) Feynman continued with his story.

"The next day at the meeting, I saw Slotnick and said, 'Slotnick, I worked it out last night; I wanted to see if I got the same answers you do. I got a different answer for each coupling - but I would like to check in detail with you because I want to make sure of my methods.' And he said, 'What do you mean you worked it out last night, it took me six months!' And, when we compared the answers he looked at mine and he asked, 'What is that Q in there, that variable Q ?' I said, 'That's the momentum transferred by the electron, the electron deflected by different angles.' 'Oh,' he said, 'no, I only have the limiting value as Q approaches zero, the forward scattering.' Well, it was easy enough to just substitute Q equals zero in my form and then I got the same answers as he did."

The following day Feynman stood up at the end of Case's talk and declared, "Your theorem must be wrong. I checked Slotnick's calculation last night and I agree with his result."

In the early days of Quantum Electrodynamics (QED) a Japanese physicist, Z. Koba, computed the very complicated corrections due to QED as well as c-meson theory in the scattering of electrons by a Coulomb potential. Unfortunately he made an error and in true samurai fashion shaved his head to acknowledge his mistake.

Dirac, who had started this whole business, wanted none of this renormalization stuff. As early as 1951 he wrote, "Recent work by Lamb, Schwinger, Feynman, and others has been very successful ... but, the resulting theory is an ugly and incomplete one". In 1975 at a lecture in New Zealand he again voiced his dislike. "I must say that I am very dissatisfied with the situation, because this so-called 'good theory' does involve neglecting infinities which appear in the equations, neglecting them in an arbitrary way. This is just not sensible mathematics. Sensible mathematics involves neglecting a quantity when it turns out to be small - not neglecting it just because it is infinitely great and you do not want it."

Actually Feynman was also not satisfied with renormalization theory and as late as 1985 he stated, "But no matter how clever the word, it is what I call a dippy process! Having to resort to such hocus pocus has prevented us from proving that the theory of quantum electrodynamics is mathematically self consistent. ... I suspect that renormalisation is not mathematically legitimate."

Dirac repeated his misgivings in 1984 at the Loyola University Symposium. "The rules of renormalization give surprisingly, excessively good agreement with experiments. Most physicists say that these working rules are, therefore, correct. I feel that is not an adequate reason. Just because the results happen to be in agreement with observation does not prove that one's theory is correct."

Dirac died on October 20, 1984 at the age of 82. That year he had published his last paper "The inadequacies of quantum field theory". In this paper he again repeated his misgivings about renormalization theory. "These rules of renormalization give surprisingly, excessively good agreement with experiment. Most physicists say that these working rules are, therefore, correct. I feel that is not an adequate reason. Just because the results happen to be in agreement with observation does not prove that one's theory is correct." He concluded the paper with, "I have spent many years searching for a Hamiltonian to bring into the theory and have not yet found it. I shall continue to work on it as long as I can, and other people, I hope, will follow along such lines."

Ten years after the Pocono conference, Bohr and Feynman met again at the Solvay conference where Feynman gave a beautiful talk on QED and renormalization theory and its experimental verification, especially with the Lamb shift. Bohr was very unhappy and asked, "Are there then no paradoxes left in QED? If so, that's terrible because then we have no way to search for a better theory."

Feynman's attitude to physics may be appreciated from the following remarks, "Physics is like sex: sure, it may give some practical results, but that's not why we do it." Also, "Nature uses only the longest threads to weave her patterns, so that each small piece of her fabric reveals the organization of the entire tapestry."

As far as quantum mechanics is concerned, Feynman had this to say. "A philosopher once said, 'It is necessary for the very existence of science that the same conditions always produce the same results.' Well, they don't!"

At Cornell, Feynman, in a lecture "The Character of Physical Law", repeated in a similar vein. "I think I can safely say that nobody understands quantum mechanics. So do not take the lecture too seriously, feeling that you really have to understand it in terms of some model that I am going to describe, but just relax and enjoy it. I am going to tell you what nature behaves like. If you will simply admit that maybe she does behave like this, you will find her a delightful, entrancing thing. Do not keep saying to yourself, if you can possibly avoid it, 'But how can it be like that?' because you will go 'down the drain' into a blind alley from which nobody has yet escaped. Nobody knows how it can be like that."

In a book with the same title, *Character of Physical Law* [2], he had this to say, "Every one of our laws is a purely mathematical statement. ... It gets more and more abstruse as we go on. Why? I have not the slightest idea."

The famous Rosenbluth formula for electron-nucleon scattering was first presented at one of the Shelter Island Conferences. It had taken M. N. Rosenbluth two years, as the major part of his doctoral thesis, to derive this result. The story goes that, on hearing of this result at the conference, Feynman wanted to verify it. So, he went to his room and reproduced the same result overnight. This does not detract from the merit of Rosenbluth, but does illustrate the extraordinary abilities of a Feynman as well as the power of his way of calculating.

In his biography Feynman talks about his discovery in 1957 of the so-called V–A theory. "It was the first time, and the only time in my career that I knew a law of nature that nobody else knew. ... It's the only time I discovered a new law." Here "V" refers to "vector" and "A" to "axial vector". Interesting is the fact that for a while experiments showed that the theory should be V+A. It was only some time after the publication of the V–A theory that experimentalists found that their earlier experiments had been in error.

There is a story that made the rounds in Caltech that when Feynman gave a talk on V–A theory he failed to mention his co-author Murray Gell-Mann (1929 –). After the talk someone is supposed to have asked him, "Professor Feynman isn't it usual to acknowledge the name of your collaborator when giving a talk on joint work?" Feynman is supposed to have replied, "Yes, but it's usual for the collaborator to have done some of the work." When Anthony G. J. Hey asked Feynman whether there was anything to this story Feynman smiled and replied, "Surely you don't believe I would do anything like that!"

At the 1979 conference on Quantum Chromodynamics, Feynman asked a sker who had used the term "Feynman integral", "What is a Feynman integral?"

As a teacher Feynman gave the following advice to the 1974 graduating class at Caltech, " The first principle is that you must not fool yourself — and you are the easiest person to fool."

In 1980 in Lisbon Feynman said he wanted to understand quark confine-

---

[2]MIT Press Cambridge MA 1985, page 39.

ment his way, so he started with the simplest non-Abelian group SU(2). "If I understand SU(2), I'll do SU(3) and I have tenure. So I have lots of time."

The story is that Feynman invented Feynman diagrams because he did not understand Second Quantization. Yasushi Takahashi (1924 –) told me that in 1953 he asked Feynman whether this was true. The answer was, "Yes."

A set of truly useful equations in quantum field theory, first derived for quantum electrodynamics, are the "Ward-Takahashi" relations. At one point Yasushi Takahashi told me, "Although I derived the relations I did not understand what they meant." He was still a student when he derived these relations. Much later, a similar set of equations was derived for other gauge theories and are known as the "Slavnov-Taylor" relations.

Takahashi has frequently told me that problems in physics are never really solved. According to him they are like a jacket with one sleeve too short. Thus, if you pull on the short sleeve to correct it, you simply pull up the other sleeve and make it too short.

In 1973 (as reported by D. Perkins twenty years later) at the Hawaii Conference on the "Development of QED" Feynman said, "When everything agrees in experiment and theory you are learning nothing. Only when things are crazy and nothing seems to work, then you are learning something."

Kip Thorne had a very good student, Bill Press and wanted to get him appointed to a three year visiting position at Caltech. To illustrate the quality of the candidate he pointed out that Bill Press had excellent physical intuition, comparable to that of Feynman. Murray Gell-Mann, always careful that Caltech's high standards should be maintained, asked, "Well, is his intuition as good as Feyman's or isn't it?" Feynman then turned to Gell-Mann and said, "Murray, if they had asked that question about you, you probably wouldn't have gotten this position."

Feynman developed his path integral method as part of his doctoral thesis at Princeton. Apparently his work with Wheeler on the absorber theory of radiation was judged by the latter to be too much of a collaboration to qualify for a thesis. Thus, Feynman turned to his interest in Dirac's work on the role of the action in quantum mechanics and the result was his "sum over paths" formulation of quantum mechanics.

Feynman loved to investigate any new problem by "turning it around". This meant that he tried for an unconventional way of looking at the problem. This passion for the unconventional sometimes was carried to an extreme and even influenced his actions outside of physics. For example, he once voted for an extreme candidate for a state office. Realizing his mistake he later sheepishly approached his colleague Murray Gell-Mann, confessed his error, and asked him in future to check his preferred list of candidates just to make sure that he was not going off the deep end.

In his later years, partly due to ill health, Feynman hated to travel and declined invitations to other institutions as a matter of course. This fact was well known. Thus, the physics students at the University of British Columbia realized that it would be a real feather in their cap and make for an outstanding event if they could get Richard Feynman to give a talk when they hosted the *Canadian Undergraduate Physics Conference*. It was a less well-known fact that Feynman had difficulty refusing a request by an attractive young lady. Whether by good luck or good judgement the students decided not to simply write to Feynman but to send a representative to invite him. The representative happened to be an extremely intelligent and also extremely attractive young lady. Feynman accepted the invitation and, in fact, returned to UBC more than once.

Like Dirac, Feynman also wanted to refuse the Nobel Prize to avoid all the publicity associated with this award. Also like Dirac he was convinced to accept the prize in order to avoid the even greater notoriety and publicity associated with refusing the prize. During his Nobel speech, Feynman confessed, "We have a habit in writing articles published in scientific journals to make the work as finished as possible, to cover up all the tracks, to not worry about the blind alleys or describe how you had the wrong idea first, and so on. So there isn't any place to publish, in a dignified manner, what you actually did in order to get to do the work."

In an interview with *Omni* Feynman had this to say, "As long as it looks like the way things are built with wheels within wheels, then you are looking for the innermost wheel — but it might not be that way, in which case you are looking for whatever the hell you find."

Feynman also had a good appreciation for the size of numbers as the following statement illustrates. "There are 10^{11} stars in the galaxy. That used to be a huge number. But it's only a hundred billion. It's less than the national deficit! We used to call them astronomical numbers. Now we should call them economical numbers."

Dirac and Feynman met for the first time at the 1961 Solvay Conference. According to A. Salam (1926 – 1996), who was sitting beside Dirac at one of the long tables around which the participants of these conferences sat, Feynman sat down across from Dirac and extending his hand introduced himself with the words, "I am Feynman." Dirac extended his hand and replied, "I am Dirac." After an uncharacteristically long silence for him, Feynman remarked somewhat awed, "It must have felt good to have invented that equation." Dirac said, "That was a long time ago." Again a long silence before Dirac asked Feynman, "What are you yourself working on?" When Feynman replied, "Meson theories," Dirac continued, "Are you trying to invent a similar equation?" Feynman said, "That would be very difficult." To which Dirac said in an anxious voice, "But one must try."

In Trieste at the International Centre for Theoretical Physics, which was started by Salam, there is a set of stairs along a steep slope, "The Scala Dirac" or Dirac Stairs, from the Centre to the park. These stairs were built because Dirac was seen attempting to climb down that steep slope and unable to do so had slid down. To prevent an unfortunate accident, these stairs were then built and named after him.

When parity violation was discovered and someone asked Dirac what he thought of that he replied, "I never said anything about it in my book." On another occasion when Abraham Pais asked Dirac why he did not mention parity or time reversal in his book, he received the following reply, "Because I do not believe in them." Already in 1949 Dirac had written, "I do not believe there is any need for physical laws to be invariant under these reflections, although all the exact laws of nature so far known do have this invariance."

This is in sharp contrast to Pauli who rejected the Weyl equations for a massless particle, out of hand, because they were not invariant under parity.

According to E. C. G. Sudarshan, "Everybody who knew Pauli knows that it was better to agree than to disagree with him". He also claimed that de Wet produced the proof of Spin and Statistics before Pauli and was criticized by Pauli.

Professor Gian Carlo Wick is remembered for a way to arrange products of field operators. This is called "Wick ordering". He also co-authored with E. Wigner and A. S. Wightman the paper that introduced superselection rules into quantum field theory. When Professor Wick arrived at Bur-sur-Yvette he was met at the train by Professor Louis Michelle to avoid what usually happened, namely that Wick got lost. Wick turned to Michelle with the comment, "Have you noticed how everyone here speaks French? Even your accent has become French." Later at lunch with the director of the institute Wick responded to a statement by the director and pointed to Michelle, "Yes, as Professor Jost over here has just explained".

Res Jost has collected the following quotes due to Kurt Friedrichs (1901 –

1982) pertaining to QFT. "Like archeologists stumbling across hieroglyphs, there is clear evidence of intelligence, but the syntax is obscure." He then continued, regarding the so-called "axiomatic" approach of A. S. Wightman: "There was no disagreement with fundamental assumptions; he tried to clear up the syntax."

In the late 1960s Klaus Hepp (1936 –) put renormalization theory on a more acceptable footing. Later, while lecturing at Brandeis University he stated, "All quantum field theory is beautiful in an empty sort of way."

In spite of all these hurdles, physicists did not lose their sense of humour. Steve Moskowsky published a paper co-authored by his cat. The editor was not impressed, but Moskowsky justified his action by claiming that the cat had contributed as much as some of his other co-authors.

Parallel to the development of QED, there were other approaches. One of these by Dirac and also later by Schwinger was an attempt to understand why charge is quantized. To achieve this, Dirac introduced the concept of a magnetic monopole. The existence of a single magnetic monopole would imply charge quantization. This was the first example of a topological quantum number. Later Dirac said with reference to his work on magnetic monopoles, "I always felt that this work of mine was somewhat of a disappointment."

In 1983 there was a conference on magnetic monopoles to which Dirac, as the founder of this subject, was invited. His response was, "I'm sorry, I am not interested in magnetic monopoles and can not come." A similar thing occurred with Schwinger, who had also come up with the idea of magnetic monopoles. Sidney Coleman met Schwinger in a parking lot in Switzerland and asked about magnetic monopoles. Schwinger replied, "At least I was smart enough to get out of this business fifteen years ago."

On Valentine's Day, Sheldon Glashow (1932 –) sent the following telegram to Blas Cabrera, who had just claimed to have experimentally discovered a magnetic monopole.

Roses are red
Violets are blue
Where is monopole number two?

Before renormalization theory, when infinities were still plaguing all quantum field theories, various sorts of remedies were attempted. One of these was to introduce a fundamental length into physics. In 1930, when he was trying to introduce a fundamental length into physics, Heisenberg wrote a letter to his former student Rudolf Peierls to get help from Pauli on non-commutative coordinates. Pauli later asked Oppenheimer who eventually asked his student Snyder. By now almost ten years had passed.

After renormalization theory's success in QED, enthusiasm for quantum field theory grew until the same techniques were applied to meson theory and failed miserably. Again many physical results, when computed using quantum field theory, turned out to be infinite. This time the renormalization theory that had worked so well for quantum electrodynamics worked not at all. Many physicists were again ready to declare field theory dead.

Heisenberg did not give up on field theory, but tried a totally new approach. All particles were to be the result of excitations of a single primary field. His theory became known as the Nonlinear Spinor Theory. Pauli was skeptical to begin with since the theory required what was then an unusual mathematical quantity, an indefinite metric. When Heisenberg asked Pauli, at a conference at Wolfratshausen in the 1950s, what the problem was, the latter explained that he had done a computation with the Lee Model and found a problem. After he explained the difficulty that he had encountered, Heisenberg left the group and went upstairs to his room. Later he returned with a smile on his face. "This is wonderful! I now have an example of what I was saying; the Lee Model works." Then Heisenberg explained what he had found.

Pauli seemed upset and in turn retired to his room. In the morning when he returned, he was very enthusiastic. Now the two of them set to work and came up with a version of an equation that might serve their purpose as a *Weltgleichung* (world equation). After the conference Pauli had to leave for the USA. Heisenberg tried to persuade him to stay since they were in the midst of this work that looked so promising. However, all arrangements for Pauli's trip had been made and so he left.

At Columbia University Pauli gave a talk on his work with Heisenberg on the nonlinear spinor theory. There were many, many questions and criticisms, not all friendly. Some of the criticism was not at all objective. Some suggested that Heisenberg was nothing more than a paranoid Nazi since he suggested that besides the proton, neutron and pions there should be a whole spectrum of particles — obvious nonsense.

In the end Pauli in most un-Pauli like fashion said, "Perhaps, my idea is crazy." Bohr, who also happened to be visiting Columbia, stood up walked a few steps to Pauli, looked down, pointed a finger at him and said, "The problem is not whether your idea is crazy, but whether it is crazy enough."

Back in Germany, Heisenberg continued to work hard. Somehow the newspapers got wind that something big was up at his institute and started snooping. Heisenberg, instructed all of his students and collaborators to say nothing since the press was sure to distort things to sensationalize them. In spite of his admonition, a few days later headlines appeared announcing that Heisenberg had discovered a new mathematics and a world equation. This was picked up by the American press and Pauli, who had been bravely defending this theory, woke up in Berkeley to just such headlines and a lengthy article extolling Heisenberg's work without a mention of Pauli's name. Pauli was clearly upset and dropped the theory.

Meanwhile Heisenberg held a press conference and set the matter straight. Too late! When Pauli returned to Europe he wanted nothing to do with Heisenberg. Heisenberg would not let it go at that and intercepted him at a meeting, "Now, Pauli, what is the trouble?"

"I don't want to talk about it."

"But I do. So, let's go to the cafeteria and you have a glass of wine and after that one can always talk better with you." At this point Heisenberg started to literally drag Pauli in the direction of the cafeteria. Pauli spotted Feynman and called, "Hey Richard, help me. Heisenberg is trying to hypnotize me."

All three of them set off for the cafeteria where Heisenberg drew out of Pauli what was bothering him about the theory. Pauli presented the difficulty that was the source of his reservation. Heisenberg replied, "So, that's what's bothering you. You know, that also bothered me for quite a while, but it's not really serious." Then Heisenberg explained a way around the difficulty. Pauli could not find a flaw in Heisenberg's argument and turned to Feynman for help. Feynman was equally unable to find a flaw. Thus, one by one Heisenberg refuted all of Pauli's arguments.

Somehow, this still did not satisfy Pauli for at the next session when, as chairman of the seminar on *New Discoveries in Physics*, Pauli introduced the first speaker he stated, "Let me begin by stating that there are no new discoveries in physics. The first speaker is Werner Heisenberg." Clearly, Pauli had totally rejected the Nonlinear Spinor Theory. In a letter to Gamow in 1958, Pauli sarcastically drew a picture to illustrate his view of the Nonlinear Spinor Theory. It was headed by the words, "Comment on Heisenberg's radio advertisement, 'This is to show the world that I can paint like Titian.'" The "painting" was simply a picture frame enclosing a blank space below which was written, "Only technical details are missing."

Until his death Heisenberg continued to work on this theory. He once explained to the philosopher, Mario Bunge (1919 –) that, "My motivation and heuristic clues have always been of a philosophical nature." He also voiced his displeasure with the excessive phenomenology of the new generation of physicists who wanted to "go back to Ptolemy". "If predictive power were indeed the only criterion for truth, Ptolemy's astronomy would be no worse than Newton's. They only wish to describe and predict facts. On the other hand I am of a Newtonian cast of mind. I want to understand facts. Therefore I appreciate the theories that explain the working of things, more than any phenomenological theories."

On another occasion Heisenberg stated: "I believe that certain erroneous developments in particle theory — and I am afraid that such developments do exist — are caused by a misconception by some physicists that it is possible to avoid philosophical arguments altogether. Starting with poor philosophy, they pose the wrong questions. It is only a slight exaggeration to say that good physics has at times been spoiled by poor philosophy."

According to Dirk ter Haar (1919 –), at a conference on quantum field theory, Pauli was asked to comment on the invited talks. His comments were, "Dirac's is classical nonsense, Heisenberg's is romantic nonsense and Heitler's is pure nonsense."

In 1976, when Heisenberg died, I happened to be in his institute in Munich on sabbatical. His student, Hans Peter Dürr was still, after more than twenty years, working on the Nonlinear Spinor Theory. As a last gesture Heisenberg gave Dürr his gold stickpin which is in the form of $\hbar$, the symbol for the quantum of action, so that he might continue to carry on the good work. Unfortunately, the Nonlinear Spinor Theory has never succeeded. In the same year Heisenberg told Karl von Weizsäcker, "If someone were to say that I had not been a Christian, he would be wrong. But if someone were to say I had been a Christian he would be saying too much".

Another approach that was tried in the 1950s was to use non-local theories. In 1952 at a conference in Copenhagen, when these non-local theories were in vogue, Pauli questioned whether a hamiltonian for such a theory would exist. A young physicist, Arthur S. Wightman (1922 –), stood up and said that by Stone's Theorem a hamiltonian must exist since the evolution is unitary. Pauli said, "I object". Wightman replied, "Now that you have objected I can continue". He then explained what he meant. Several months later, a hamiltonian was found and Pauli wrote a letter to Wightman as a "peace-offering from a cannibal to a missionary".

Wightman went on to found what came to be known as Axiomatic Quantum Field Theory by all but its practitioners, who referred to it as Constructive Field Theory. His idea was that "If we can't be sure that the physics is correct, we can at least make sure that the mathematics is correct".

When Andrei Grib visited Edmonton in 2001, he related the following. On his visit to Princeton he met Arthur Wightman who, as they passed Edward Witten's (1951 –) office, pointed to it and said, "There is Ed Witten. He does not use quantum mechanics. He only uses functional integrals which do not exist and he proves all sorts of results with them, some of which are quite interesting." I find this story somewhat hard to believe since I once heard Wightman stop a discussion when one of the disputants tried to hide behind some abstruse mathematical theorems. "Don't hide behind the formalism: stick to the physics."

After the failure of renormalization theory when applied to mesons, attempts were made to replace it with dispersion relations which had been, in a sense, indicated by field theory. Marvin Goldberger, one of the champions of dispersion relations, stated quite bluntly, "Any attempts to derive dispersion relations from field theory are like tits on a man, neither useful, nor decorative". I also remember that as a graduate student in one of his lectures he came in one day and told a story the gist of which was as follows.

Sometime in the future, after dispersion relations have completely replaced all of quantum mechanics and quantum field theory, a bright young graduate student will show his professor a wave equation that he has found. This equation, just like the dispersion relations, also allows one to calculate the amplitude for a given process. The old professor looks at the equation and after a while tells the student, "Yes, this equation used to be known as the Schrödinger equation."

These attempts, at using dispersion relations, lasted for a while but proved inadequate. Eventually field theory proved to be more robust than any of these other approaches. Finally in the seventies the richness of gauge field theories led to a unification of the electromagnetic and weak interaction. This was accomplished mainly at the hands of Sheldon Lee Glashow (1932 –), Abdus Salam (1926 –), and Steven Weinberg (1933 –). Essential for their work was a result by Peter Higgs (1930 –) that made it possible for massive vector theories to look like QED and become renormalizable. To do this it was necessary for a new kind of particle, the Higgs particle (so far undiscovered) to exist.

The Higgs model introduces a parameter with the dimensions of mass into the theory. The theory starts out describing a vector particle without mass and then, the vacuum is such that a scalar field assumes a value in the vaccum state. This is the the Higgs mechanism and the vacuum value allows the vector field to describe a massive particle. This has led some physicists to claim that the Higgs mechanism explains the origin of mass. However, the Higgs mechanism has only shifted the origin of mass from one parameter to another. The real purpose of the Higgs mechanism is to render renormalizable a theory that is not renormalizable to begin with.

Frank Wilczek (1951 –), winner of the 2004 Nobel Prize in Physics in his article *The Origin of Mass* has this to say. "...the masses of W and Z bosons, and of quarks and leptons, arise from the interaction of these particles with the pervasive Higgs condensate. This has inspired references to the Higgs particle as 'the origin of mass', or even 'the God particle'. The real situation is interesting but rather different from what this hyperbole suggests. ...for quarks and leptons the Higgs mechanism appears more as an accommodation of mass than an explanation of its origin." he continues with the statement, "...the Higgs field in no sense explains the origin of its own mass. A parameter directly equivalent to that mass must be introduced into the equations explicitly."

After Don Perkins asked Peter Higgs whether he had any doubts about the particle named after him, Higgs approached Dirac the next day at tea. "Mr. Dirac, did you ever have any worries about your equation?" According to Don Perkins the response was, "No! I never had worries. I wrote an equation for how the real world could be. If the real world wanted to be different, so much the worse for it."

Also according to Don Perkins, Salam wanted his name kept off the Salam, Ward, Weinberg, Glashow angle since he thought that it was "crap". However, after Salam got the Nobel Prize, he no longer remembered that conversation.

A very exciting idea, now very much in vogue in quantum field theory, is the idea of supersymmetry. Around 1940 Pauli seems to have anticipated this idea, but rejected it. He had stated, "If the number of fermions equals the number of bosons in the theory then the positive vacuum energy of the bosons will be cancelled by the negative vacuum energy of the fermions, but nature is *not* like that."

Supersymmetry is a theory that requires nature to be like that. In other words, supersymmetry requires that for every fermion there has to be a corresponding boson partner and vice versa. Not a single one of these supersymmetric partners has been detected so far. This has led to the following sarcastic statement, "Supersymmetry has experimental evidence since half the required particles have already been detected." At the 2000 Higgs Fest in Edinburgh to celebrate his seventieth birthday Peter Higgs wore a T-shirt with a picture of his grandson thus showing the existence of "a second light Higgs". This was "considered" by some people as "possible evidence of supersymmetry".

Julius Wess, one of the inventors of supersymmetry has this to say, "The number of years required to verify a result is proportional to the number of papers published. For supersymmetry we have about 30, 000 papers in the first 30 years."

At present field theory is again trying to nurture new off springs: string theories. These theories are now approaching middle age without any experimentally verifiable predictions to date. They aim at nothing less than a unification of all the known interactions. The jury is still out on whether this program will succeed, but in any case, quantum field theory is alive and well.

Not everyone is sure that string theory is the answer. In an interview[3], Roger Penrose (1931 –) had this to say, "String Theory is an example of science being driven by fashion. And I have mixed feelings about it. Some of the mathematical notions which people associate with string theory are very appealing. But, just because they are appealing, doesn't mean that they are right."

Similarly, Sheldon Glashow (1932 –) stated in a more sarcastic vein, "Contemplation of superstrings may evolve into an activity as remote from conventional particle physics as particle physics is from chemistry, to be conducted at schools of divinity by future equivalents of mediaeval theologians."

It is interesting to contrast this with the statement by Edward Witten (1951 –). "Good wrong ideas are extremely scarce, and good wrong ideas that even remotely rival the majesty of string theory have never been seen before."

That even very good physicists are prone to outrageous exaggeration is illustrated by the following statement due to Stephen Hawking [4], "... by the end of the century ... we might have a complete, consistent, and

---

[3] *New York Times*, Jan 19, 1999, page D3
[4] *Physics Bulletin*, *32*, 15, (1981)

unified theory of the physical interactions which would describe all possible observations."

The particle theorist Frank Wilczek from the Institute for Advanced Study in Princeton made the following comment, "In physics you don't have to go around making trouble for yourself — nature does it for you". He also has asked, "To what extent are the laws of physics unique?" He then went on to outline three possibilities. "The first option is that there is a unique fundamental equation with a unique stable solution. The second is a unique, fundamental equation that has many consistent solutions, with the particular solution that describes our world being picked out by some historical accident. Perhaps different, distant parts of the universe, not yet observed, experience different laws. Thirdly, there are a range of possible equations, each of which is equivalent for 'practical' purposes and for which the first two options apply."

Chapter 5

The Strongest Force

"In order to account for this large angle scattering of α particles, I supposed that the atom consisted of a positively charged nucleus of small dimensions in which practically all the mass of the atom was concentrated." E. Rutherford, Philosophical Magazine, 27, 488, (1914).

In 1898, two years after Becquerel discovered radioactivity, Pierre Curie (1859 – 1906) and Marie Sklodowska-Curie (1867 – 1934), together with Gustave Bémont (1857 – 19320 published a paper in *Compte Rendu, Paris*, *127*, 1215, (1898). This paper dealt with radiation emanating from radium. There they stated, *"On réalise ainsi une source de lumière, a vrai dire très faible, mais qui fonctionne sans source d'énergie. Il y a là une contradiction tout au moins apparent avec le principe de Carnot."* (One realizes thus a source of light, to speak the truth very weak, but which functions without a source of energy. There is here a contradiction, or so it seems, with the principle of Carnot.). The "principle of Carnot" is the French name for energy conservation. What this paper referred to was the puzzling fact that energy emanated from radium without any measurable chemical changes.

The puzzle was solved by E. Rutherford and F. Soddy, *Philosophical Magazine*, *4*, 370, 569, (1902). They proposed that radioactive materials contain unstable atoms of which a fraction decay into new unstable atoms and so on until a stable atom is reached. They called the process "transformation theory" . According to M. Howorth, [1] the following exchange took place

---

[1] *The Life of Frederick Soddy*, New World Publishers (1958)

between Rutherford and Soddy.

Soddy: "Rutherford, this is transmutation ..." .

Rutherford: "For Mike's sake, Soddy, don't call it transmutation. They'll have our heads off as alchemists."

In his 1904 book, *Radioactivity*, Rutherford wrote with regard to transformation theory. "This theory is found to account in a satisfactory way for all the known facts of radioactivity, and a mass of disconnected facts into one homogeneous whole. On this view, the continuous emission of energy from the active bodies is derived from the internal energy inherent in the atom, and does not in any way contradict the law of conservation of energy."

Physicist also found that there were three kinds of radiation which Rutherford labelled: alpha, beta, and gamma. He later identified alpha radiation as Helium nuclei. J. J. Thomson identified beta radiation as electrons, and gamma radiation was found to consist of very high energy X-rays. There was one more particle that was known to exist, the nucleus of the hydrogen atom, the proton. These then were the constituents of nature as understood at the beginning of the twentieth century.

The proton was found to also have spin and this was discovered in a most extraordinary way. Friedrich Hund (1896 – 1997) had quit his position in Leipzig as Heisenberg's assistant and gone to Copenhagen. Gossip has it that part of the reason for his departure was due to his name. While working with Heisenberg, all their papers were published in German with the authors listed alphabetically as *"Heisenberg und Hund"*. Since in German all nouns are capitalized, this also translates as "Heisenberg and dog". Furthermore, the lectures at Leipzig listed *Heisenberg mit Hund.* This could be read as either Heisenberg with Hund or Heisenberg with dog.

Early in 1927, Hund came up with the idea of looking at the band spectrum of the hydrogen molecule to establish whether the proton is a boson or a fermion and what the spin of the proton might be. However, not enough data were available, so he decided to use the known data for the part of the hydrogen molecule specific heat due to rotation. His careful computation did not lead to a definite conclusion although he said that the best fit to the experimental data occurred if one assumed that the proton was a boson with spin $1/2$.

Hund had done all this work on the band spectrum of the hydrogen molecule at Bohr's institute in Copenhagen. Shortly after he published this result, a Japanese expert in spectroscopy, Takeo Hori (born 1899), arrived there. With the pressing need to clarify the spectrum of the hydrogen molecule, Bohr suggested that Hori study this problem. After an elaborate experiment, Hori came up with very clean data that disagreed with the results that Hund had obtained. The difficulty was resolved by D. M. Dennison who realized that Hori's data implied that certain states in the hydrogen molecule were so stable that equilibrium between the various states could not occur in the short times that it took to measure the specific heats. But equilibrium is precisely what Hund had assumed in his calculation since there was no reason to suspect that things were otherwise. This explained the discrepancy. Not only that, but the ratios of the intensities of the lines also lead to the conclusion that the proton was a fermion, not a boson, and that its spin was $1/2$.

Rutherford had been one of the first to measure the energy released in radioactive disintegrations and found that this energy was enormously large. At this time there was a huge discrepancy between the geologists who claimed a very old earth and Lord Kelvin who had calculated the time for a molten sphere to cool to the present interior temperature of the earth and had thus found a much shorter age. Rutherford thought that radioactive energy released inside the earth could solve the problem and after performing a calculation was convinced that this was so. He was very proud of this result. However, when he went to present his findings, Lord Kelvin was in the audience.

Kelvin was a preeminent figure in physics and his estimate of the age of the earth on the basis of how long it would have taken the earth to cool to its present temperature had to be taken seriously. Even Darwin, who had originally stated that the time for evolution to occur had to be much longer than Kelvin's estimate of 100 million years dropped this statement from the second edition of his *The Origin of Species* in order not to come up against Kelvin. Clearly Rutherford had cause to worry about Kelvin's response. He described the encounter thus.

"I came into the room, which was half dark, and presently spotted Lord Kelvin in the audience and realized that I was in for trouble at the last part of my speech dealing with the age of the earth, where my views conflicted with his. To my relief, Kelvin fell fast asleep, but as I came to the important point, I saw the old bird sit up, open an eye and cock a baleful glance at me! Then a sudden inspiration came, and I said Lord Kelvin had limited the age of the earth, provided no new source was discovered. That prophetic utterance refers to what we are considering tonight, radium! Behold! The old boy beamed upon me."

Rutherford was very much the no nonsense experimentalist. He once told a student, "All science is either physics or stamp collecting." On another occasion he engaged in a debate with a well-known philosopher and exclaimed that all of philosophy was nothing but hot air. The philosopher countered by stating that Rutherford was a savage, a noble savage, but still a savage. The philosopher then continued with a story of Marshall McMahon of Napoleon III. "While reviewing a regiment in which there was a Negro cadet, the Marshall had been asked to say something encouraging to him. When he reached the cadet's platoon, the Marshall stopped, looked at him and said, 'Cadet, you are a Negro.' The cadet replied, 'Yes, sir!' After a pause the Marshall continued, 'Well, go on being one.' Well, Rutherford, that's what I say to you, go on."

In the middle of the night after a hard day in the lab, the telephone rang in Mark Oliphant's home. His wife answered to hear Rutherford on the line. Her sleepy husband heard his boss, Lord Rutherford shout into the phone, "I know what the particles are. They are Helium of mass 3." Oliphant immediately agreed, "Yes, sir. But why do you think they are Helium 3." The response was typical, "Reasons, reasons? I feel it in my water!" Rutherford's bladder was right.

That even a master such as Rutherford could be mistaken in the abilities of people was illustrated more than once. While working with Ramsay, the radiochemist, Otto Hahn (1879 – 1968), not only discovered that there were at least two isotopes of radium, but also found another isotope of thorium, Th 228, which he dubbed "radiothorium". The top radiochemist at Yale, Bertram Borden Boltwood (1870 – 1927) did not believe these results and wrote to his friend Rutherford that radiothorium was "a compound of thorium and stupidity". Thus, when Hahn arrived at Montreal, Rutherford was not at all convinced of his abilities. However, soon after Hahn arrived at McGill he was able to convince both Rutherford and Boltwood that he was right.

Boltwood later achieved fame by realizing that lead might be the end-product of the radioactive decay of radium and thorium. He then used this idea to measure the age of the earth and came up with 2.2 billion years, much older than even geologists had previously thought.

Hahn remained with Rutherford from 1905 to 1906. He then returned to Germany. In 1907 he teamed up with the Viennese physicist, Lise Meitner (1878 – 1968) and continued to work together with her until 1938. Their collaboration eventually lead to the atom bomb.

Another one of Rutherford's collaborators for several years was Frederick
Soddy (1877 – 1956). This came about as follows. Soddy had applied for
a position in Toronto. When, after some time he had no reply, he took a
boat from England to New York. On arrival there he learned that the post
in Toronto had already been filled, but decided to finish his trip in Montreal
where he presented his letters of reference and was hired in the department of
chemistry. Here, he and Rutherford became close friends and collaborators.
This is where they established the theory of radioactive transmutation of the
elements. Soddy also invented the term "isotope".

Another example of Rutherford's inability to predict future research was
his statement with regard to cosmic rays. He announced to the London International Conference in 1934, "The swiftest particles known to accompany
the cosmic rays have energies of the order of 100 million volts ... some are
believed to have energies even greater than 1,000 million volts. ... Information of this kind is of great value, for it is unlikely that we shall ever be
able to produce particles of such energy in the laboratory." No wonder he
had in earlier years shouted at James Chadwick (1891–1974), "I won't have
a cyclotron in my laboratory."

In spite of this John D. Cockroft and Ernest T. S. Walton not only built
a cyclotron, but also used it to split the Lithium atom. This was the first
case of artificially induced nuclear fission. After this experiment Cockroft is
reported to have walked the streets of Cambridge and addressed strangers
with the words, "We have split the atom! We have split the atom!".

In the same scattering experiments in 1911, in which Rutherford had
found that atoms consisted mostly of empty space with a tiny nucleus at
the center, he also found that this nucleus was about 100, 000 times smaller
than an atom. Yet in this tiny space was a positive charge as big as the
negative charge of the electrons in the atom. This meant that a new force,
much stronger than the electromagnetic and of much shorter range held the
nucleus together. This was the beginning of nuclear physics as a field of
research in its own right. The nature of this force, now known as the "strong
force", was a source of experimental investigation for at least half a century
and still has some problems associated with it.

In spite of the very strong nuclear force, the amount of energy that could
be extracted in the laboratory was miniscule. This lead to the following, soon-
to-be-overthrown, prediction by Lord Rutherford in an address to the British
Association for the Advancement of Science, in September, 1933. It happened
in the same room in which a quarter century earlier Lord Kelvin had declared

the atom to be indestructible. "The energy produced by the breaking down of the atom is a very poor kind of thing. Anyone who expects a source of power from the transformation of these atoms is talking moonshine."

The mystery of the nucleus was deepened by the fact that the force holding it together was so strong (about a thousand times stronger than the electromagnetic) and acted over only a very short distance of about 0.000 000 000 001 mm. Also, the nucleus of many of the heavier elements emitted all three kinds of radiation: alpha, beta, and gamma. All sorts of models of the nucleus were constructed, most of them wrong. Thus, as late as 1925 Sommerfeld wrote, "There can be no doubt that the radioactive nuclei contain Helium nuclei and electrons which they emit as alpha and beta rays." This very reasonable conclusion was wrong as far as the electrons are concerned. A straightforward application of Heisenberg's uncertainty relation, would have shown this, but of course this knowledge was not yet available.

Sommerfeld was not alone; most of the earlier nuclear models, until about 1933 visualized the nucleus as consisting of protons and electrons. So, even after the uncertainty relation was known, it was not applied to this problem for seven years. After Chadwick, in 1932, discovered the neutron Heisenberg proposed, in a letter to Bohr, that in spite of the fact that the neutron might be a complicated bound state of a proton and electron, it might still be possible to consider it as elementary. "The basic idea is to shove all the difficulties of principle onto the neutron and to apply quantum mechanics within the nucleus." On the other hand he was not ready to banish the electron from the neutron, even though this violated the uncertainty relation, as well as the statistics of spin $1/2$ particles. Instead, he even contemplated the necessity to sacrifice quantum mechanics. Thus, he wrote that if the neutron is a compound of a proton and an electron, "In defense of this hypothesis, one can at once adduce that the very existence of the neutron contradicts the laws of quantum mechanics in their present form."

Around 1927, Petrus Josephus Wilhelmus (Peter) Debye (1884 – 1966) had this to say with respect to the problem of electrons in the nucleus. "This is something best ignored, like the new taxes."

In 1929 George Gamow (1904 – 1968) wrote a book on nuclear physics. At the time the neutron had not yet been discovered and there was this difficult question about electrons in the nucleus. The Dirac equation showed that it was impossible to confine an electron to the volume of a nucleus. Thus, as a warning to the reader, Gamow had a special stamp, with skull and cross bones, made up and used this to mark all the passages in his book that dealt with electrons in the nucleus. The publisher, Cambridge University Press, however wanted to change all the skull and cross bones to a less frightening symbol. Gamow agreed. "It has never been my intention to scare the poor readers more than the text itself will undoubtedly do."

Gamow also was the first one to come up with the idea of quantum tunneling. Prior to this, using the concepts of classical mechanics, made it very unlikely that fusion reactions could take place in the sun. Eddington stated, "We do not argue with the critic who urges that the stars are not hot enough for this process; we tell him to go and find a hotter place ." The hotter place became unnecessary with the discovery of quantum mechanical tunneling.

George Gamow also explained the origin of alpha emission. In *Cosmology, Fusion, and Other Matters*, edited by F. Reines, Léon Rosenfeld tells the following story of this work. "In my experience nuclear physics starts with the sudden appearance, one morning in the library of the Göttingen Institute, of a fair-haired giant, with shortsighted, half-shut eyes behind his spectacles, who introduced himself with a broad smile, by declaring, 'I am Gamow'. This pronouncement, at that time, could not provoke very much excitement. As it turned out that Professor Born would not be in for some time, I proposed to Gamow to go out for a walk. It was during this walk that he told me what he was doing.

"He wanted to understand alpha radioactivity. Now, this seemed to me — and I think most physicists then would have had the same reaction — a quite fantastic idea. All we knew about nuclei was that they were very small and that they had spin. ..."

Gamow used quantum mechanics and showed that this theory predicted that particles could jump (or tunnel) from one side of an energy barrier to the other. This is called quantum tunneling. At the time the idea seemed to even contradict the mathematics of quantum theory and Born at first rejected the whole thing. However, when Gamow's idea turned out to give results in concordance with alpha decay for nuclei with lifetimes ranging from fractions of a second to millions of years, Born took the idea seriously and found that Gamow's work did not contradict anything.

When Gamow presented a talk on alpha decay and barrier penetration in England, Ralph H. Fowler (1889 – 1944) remarked, "Anyone present in this room has a finite chance of leaving it without opening the door, or of course, without being thrown out of the windows."

Incidentally, Gamow wrote several very readable books for non-physicists.

It may have been his literary style that made him prone to use hyperbole. On one occasion he stated, "The sun is much, much bigger than even an elephant."

In 1932 Born prepared a paper on the quantum theory of the nucleus. It was written in longhand and since, in the older German way of writing, n's and u's are very similar, the title of the conference, as typed by his stenographer, came out as "For the conference on unclear physics" instead of "For the conference on nuclear physics". This typo has become a standard way for other physicists to poke fun at their nuclear colleagues.

That nuclear physics became a major research area is evidenced by the fact that between 1930 and 1940 the number of funded papers on nuclear physics published in the *Physical Review* increased from a negligible proportion to 55%. No wonder that later Marvin Goldberger (1922 –) in Princeton stated with regard to nuclear physics that, "Never before in the history of physics have so few owed so little to so many." Later when Goldberger became a member of the Institute of Defense Analysis called the Jason Group he was asked where the name came from. His answer was, "It's a golden fleecing."

To return to the historical development, in 1928 Walther Bothe and his student Herbert Becker bombarded beryllium with alpha particles from polonium. They found that the emitted radiation had more energy than the incident alpha particles. This had to come from nuclear disintegration.

Walther Bothe (1891 – 1957) succeeded Philipp Lenard (1862 – 1947) in 1932 as professor at the University of Heidelberg. The physics laboratory had numerous huge electric meters all over the place. For quite a while visitors noted that when Bothe showed them around the lab he would once in a while take out a small penknife and scratch at one of the meters with it. His explanation was that Lenard had felt that the international commission that had named the unit of current the "ampere" had made a grave error. In his opinion it should have been called the "weber" after the German physicist. Thus, Lenard had little labels with the word weber pasted over the word ampere and Bothe was busy removing these labels as he discovered them.

Walter Bothe also managed to alienate Geheimrat Dr. Karl Scheel (1866 – 1936), the editor of *Zeitschrift für Physik*. Professor Rudolf Frerichs reported this to my colleague, Frank Weichman, in a letter. Frerichs was present when the following occurred. Karl Scheel was a rather rotund elderly gentleman with fleshy hands of a somewhat rosy complexion. Bothe made the following remark. "Herr Geheimrat, your hands together with a puree of peas and some sauerkraut would present a magnificent *Eisbein.*" The *Eisbein* is pig's knuckle and one of the local specialties of Berlin.

Frédéric Joliot (1900 – 1958) studied with Paul Langevin (1872 – 1946) who had tried to discourage him from physics. "A career in science is very risky. Unless you make an important discovery you will never go far." Joliot persisted and in 1925 Langevin got him a job at Marie Curie's Radium Insitute as her assistant. There he met her daughter, Irène Curie (1897 – 1956) and, the following year, married her. Some Paris scientists suspected him of marrying her to advance his career and started calling him the "prince consort". Stung by this gossip, Joliot was more determined than ever to prove himself. Maybe his mother-in-law's statement may have helped. She had written, *"En science nous devons nous intéresser aux choses, non aux personnes."* (In science we should concern ourselves with things, not with people.) In 1930, after he got his Ph.D., Joliot was unable to find a job due to the rampant nepotism prevailing in French science at that time. He even considered a career in industry.

Fortunately, about this time several of the notable scientists in France, including Jean-Baptiste Perrin (1870 – 1942) and Paul Langevin managed to get the research funding in France reorganized. So, it came about that in 1931 Joliot was one of the first beneficiaries of a research position with a salary funded by the Caisse Nationale. With the help of Perrin he got a half time position at the Radium Institute and half time elsewhere. Irène also eventually got a position paid by the Caisse Nationale.

Later, Frédéric Joliot and Irène Curie took up the study begun by Bothe and Becker. They found that the radiation produced could emit protons from paraffin. This was very surprising and they tried to interpret this (in a paper published in 1932) as some very energetic gamma radiation, a sort of Compton effect. However the energy of the gamma rays required was much too high. When Chadwick presented this result to Rutherford, the noble lord snorted, "I do not believe it". Similarly, Ettore Majorana (1906 – 1938?) [2] in Rome sneered, "What fools! They have discovered the neutral proton and do not recognize it." Indeed they had missed the Nobel Prize. Actually they missed the Nobel Prize twice. They had also, before Anderson, observed positrons in their cloud chambers, but had misinterpreted them

---

[2]The date of Majorana's death is uncertain since he disappeared on a boat trip from Palermo to Naples in March, 1938. Numerous conjectures arose: suicide, accidental drowning, retreat to a monastery, kidnapping. Since he was a brilliant scientist, the case became one of major national interest in Italy.

as electrons going in the opposite direction even though it was difficult to understand where these electrons might have come from. They finally did receive the Nobel Prize for chemistry in 1935 for being the first to synthesize new radioactive elements. Irène Curie's mother, the famous Marie Curie, wrote congratulations from her deathbed to her daughter, "We have here returned to the glorious days of the old laboratory."

James Chadwick immediately repeated the experiment of Frédéric Joliot and Irène Curie, but went further and compared the recoil obtained from collisions with helium as well as with nitrogen. In this way he was able to show that the neutral radiation consisted of a particle with approximately the same mass as the proton. Thus, he had discovered the neutron.

While working on the detection of this neutral particle, the neutron, Chadwick worked day and night. It became a standard joke at the Cavendish to ask, "Tired, Chadwick?" The response always was, "Not too tired to work." After three weeks he had the results and told his colleagues, "Now I want to be chloroformed and put to bed for a fortnight."

About this time (1935) Joliot told Elsasser that he thought that experimental nuclear physics was the type of activity that would lend itself to organization like an industrial enterprise: One man to run each type of machine, one man to run the accelerator, one to run the counters, and so on, with one man at the center to do the thinking.

During World War Two, Joliot joined the French resistance and helped hamper German efforts to develop an atomic bomb. At that time he also joined the French communist party and remained a communist to the end of his life. This led to difficulties since after the war he was made head of the French Atomic Energy Commission and France was a member of NATO. Eventually the French government succumbed to pressure and he was removed from his post and replaced by Francis Perrin. Around 1936, after Hitler had come to power, Perrin had made the following comment, "I don't understand these Germans. Why do they want to kill off science, the cow that nourishes them?"

After Chadwick's discovery of the neutron, Heisenberg almost immediately began to study the nucleus by treating the neutron and proton as two equal but distinguishable constituents of the nucleus. To distinguish them he introduced the concept of isospin. In 2001 when Jürgen Ehlers was reviewing Heisenberg's achievements at the Alexander von Humboldt symposium to celebrate the 100th anniversary of Heisenberg's birth he mentioned that in 1932 Heisenberg introduced the concept of isospin and went on to say, "This was very courageous."

Enrico Fermi (1901 – 1954) decided to use Chadwick's neutrons to get into the nucleus since, being neutral, the neutrons would experience no electric repulsion, as would protons. This required a source of neutrons, which Fermi got by bombarding beryllium with alpha particles. His group in Rome succeeded in producing new isotopes for every element above oxygen that they bombarded. Fermi tried to produce transuranic elements and seemed to succeed, since entirely new chemical elements were found, and these differed from any of the chemical elements close in the periodic table to the ones bombarded. These studies were taken up both in Berlin and Paris and many more such results were found. The new chemicals were, however not transuranic, but signalled something even more interesting.

While Fermi and his collaborates were bombarding various nuclei with neutrons, a serendipitous event occurred. In testing some metals for artificial radioactivity Bruno Pontecorvo (1913 – 1993) and Edoardo Amaldi (1908 – 1989) discovered a most strange result. The radioactivity increased when the material being irradiated by neutrons was shielded from the neutron source. They summoned Fermi. After some experimenting they found that a shield of uranium gave very little increase but a shield of wood such as the tabletop increased the radioactivity more than anything else they tried. Fermi then suggested that they use something really light such as paraffin. The Geiger counter went wild. By now it was lunch time and Fermi suggested that they go and eat. This experiment illustrates a statement by Fermi. "If you find agreement with theory you have made a measurement, but if you are very lucky you will find disagreement. Then you have made an experiment."

During lunch, Fermi came up with an explanation for these strange results. A light material such as paraffin is rich in hydrogen. The neutrons colliding with the protons (which have almost the same mass as neutrons) in hydrogen lose a lot more energy than in a collision with a heavy nucleus. The low energy neutrons then react more readily with the nucleus into which they are absorbed since they do not merely pass through with some deflection. This idea needed testing. The only cheap material rich in hydrogen is water. So off they went into the courtyard and submersed their sample for testing in Senator Corbino's fish fountain. Again the Geiger counter went wild. Fermi was right.

The problem of beta emissions presented a far more serious problem. It soon became obvious that the law of conservation of energy as well as the law of conservation of momentum seemed to be violated in beta decay. This lead to such outrageous suggestions as the one by Bohr that these laws might be valid only in a statistical sense, but not for individual processes. Einstein's reaction to this was very strong. In a letter to Born he wrote, regarding the possibility that causality or energy momentum conservation might have to be given up, "Then I would rather be a shoemaker or an employee in a gambling casino than a physicist."

Pauli also could not accept this and in a letter to Heisenberg wrote that he would "publicly confess" his belief in the conservation laws. He continued, "I shall insist in Brussels that a neutron can never decompose into an electron and a proton." He referred to Brussels because the next Solvay Conference, in Brussels, was coming up. On another occasion he stated the same thing. "We are inclined to believe that what is discovered in physics related to the conservation of energy, momentum, etc. has to be right and correct because the human mind and the objects that we see and that we discover belong to the same perennial order."

When the problem with non-conservation of energy in beta-decay was very much in the forefront, Pauli sent a letter to Bohr addressed to, "Dear Radioactive Ladies and Gentlemen." In this letter Pauli suggested that in beta decay a very light neutral particle might be emitted and that this particle carried away the energy and momentum necessary for conserving these quantities. When this letter arrived Bohr was unable to make up his mind about it. In order to gain more time to think about this problem, he asked his wife to write to Pauli with some family news and to mention that, "Niels will write on Monday." A few weeks later another letter arrived from Pauli, this time to Mrs. Bohr. In this letter Pauli wrote that Mrs. Bohr had been very wise not to mention in her letter on which Monday Niels was going to write. He also added that, "Niels should, however, in no way feel bound to write on Monday. A letter written on any other day would be just as welcome."

Several other people received Pauli's letter. He dubbed the required light neutral particle "neutron", but after Chadwick's discovery of the neutron the Italians, notably Fermi's group, renamed the particle "neutrino" meaning, "little neutral one". This came about at a seminar where Fermi discussed Chadwick's discovery of the neutron. Someone asked if it was the same as Pauli's neutron. "No," explained Fermi, "Chadwick's neutron is big and heavy. Pauli's is small and light; it would be better to call it a neutrino." For quite some time Pauli hesitated to publish his idea. Finally, a year later at the Brussels conference, he proposed it in public.

"With regard to these neutral particles, we first learn from atomic weights (of radioactive elements) that their mass cannot be much larger than that of the electron. In order to distinguish them from the heavy neutrons, Fermi proposed the name 'neutrino'. It is possible that the neutrino proper mass is equal to zero, so that it would have to propagate with the velocity of light, like photons. Nevertheless, its penetrating power would be much greater than that of a photon with the same energy. It seems to me admissible that neutrinos have a spin 1/2 and that they obey Fermi statistics, in spite of the fact that experiments do not provide us with any direct proof of this hypothesis."

Fermi almost immediately developed a theory using the neutrino to describe beta decay. This theory is now called the "Fermi interaction". He submitted this paper with all the calculations and a surprisingly good fit to experimental data to *Nature*. This journal rejected the paper as containing "abstract speculations too remote from physical reality to be of interest to the readers." This occurred before Fermi's experiments with artificial radioactivity and may have induced him to take up some experimental work as a vacation from theory. He then sent a somewhat expanded version of his theory to *Ricerca Scientifica* where it was accepted.

The details of Fermi's calculations agreed with experimental results to such an extent that the existence of the neutrino was soon accepted, even though direct detection of the neutrino did not occur until the 1950's. No wonder Felix Bloch waxed rhapsodical when he wrote to Gregor Wentzel, "Fermi has made a beautiful theory of beta-decay emission, introducing the neutrino, which so simply reproduces the empirical facts that I believe in it strongly." When Pauli heard he also was ecstatic and wrote to Heisenberg, "That would be water onto our mill!"

The same Bruno Pontecorvo mentioned above, later did excellent work on neutrino physics and was awarded the Eötvös Prize. When a lady reporter asked him if there were "any hope that neutrinos will sometime be of benefit to mankind", Pontecorvo demonstrated his sense of humour. "Why do you say 'will'?" And pointing to himself continued, "They already benefit some now."

In 1932, when beta decay was still very much discussed and, when particle creation was not yet a physics concept, it was difficult to understand how both electrons and neutrinos could emerge from the nucleus. At that time Heisenberg was a young professor in Leipzig and Victor Weisskopf was his unpaid post-doctoral assistant, supported by his parents. They, together with some students, were sitting at a coffee house across from the entrance to an indoor swimming pool and discussing precisely this question of how an electron and a neutrino could emerge from a nucleus. Heisenberg brought their attention to the door of the indoor swimming pool. "You are all drawing the wrong conclusion. Everyone you see is entering and exiting that building fully dressed. Should you conclude from this that the swimmers inside are also fully dressed?"

After Fermi had written down the Fermi interaction it was hoped that the neutrino jumping back and forth between nucleons, just like the photon jumping back and forth between charged particles, would produce the attractive force between nucleons. Unfortunately, or perhaps fortunately, a calculation in 1934 by Dmitrii Dmitrievich Ivanienko (1904 – 1994) and Igor Y. Tamm (1895 – 1971) showed that the Fermi interaction was a million million times too weak. This, as already discussed, then led Yukawa to his theory. The force described by the Fermi interaction is now called the "weak force".

Many years later, after his theory was already well-established, Fermi kept a matchbox on his desk with the standard label, "Guaranteed to contain 50 matches". He had crossed out the word matches and replaced it with neutrinos. If one computes the solar neutrino flux one finds that there should be a density of about 10 neutrinos per cubic centimeter at the earth's surface.

When Fermi was still a young professor at the University of Rome, Mussolini gave him the title *"Eccelenza"* (His Excellency"). On one occasion Mussolini was going to address the Academy of Sciences. As usual, Fermi arrived in his little Fiat only to find that everyone else was arriving in huge limousines with liveried chauffeurs. When stopped at the gate by the carabinieri and asked for his business, Fermi realized that no one would believe him if he said that he was His Excellency, Enrico Fermi. Consequently he said that he was the driver for His Excellency, Enrico Fermi.

"Very well, drive in, park over there and wait for your master."

In 1934 at a conference, Hans Bethe (1906 –) and Sir Rudolf Peierls (1907 – 1995) concluded that one would never be able to see or hear the neutrino. Many years later, at a conference to celebrate Peierls, the two colleagues were again united. By this time Bethe was 83 and had difficulty seeing, while Peierls had difficulty hearing. Thus, when someone asked a question with reference to an equation about the neutrino, Bethe had to tell Peierls what the person had said and Peierls had to tell Bethe who the questioner was.

Fermi's experiments had shown that materials such as water or paraffin, rich in hydrogen could slow neutrons down, and that these slow neutrons were very easily captured by nuclei. This occurred in 1935. Bethe then tried to calculate the capture probability and his result was reported at a colloquium in Bohr's Institute. Bohr found the result difficult to understand and kept interrupting the speaker with questions until at one point he became totally silent and his face turned slack. After a period of silence he smiled, "Now I understand it." This was the "aha!" moment at which he had understood the compound nucleus. The neutron simply shared its energy with the other neutrons and protons in the nucleus and thus did not have enough energy to escape.

Also in April 1936 Gregory Breit (1899 – 1981) and Eugene Wigner published in the *Physical Review* a paper in which they assumed that the slow neutron formed a virtual energy level with the nucleus and then decayed by gamma emission. They worked out the details of this theory and its agreement with experiment gave support to Bohr's more general theory which then dominated nuclear physics for the next 20 years.

Bohr and Fritz Kalckar (1910 – 1938) collaborated on elaborating Bohr's theory. They published their results in October 1937, the same month in which Rutherford died. They continued to work together and Bohr also elaborated the liquid drop model of the nucleus, an idea due to George Gamow. Bohr did not mention Gamow's work. So, the model is named after Bohr, since Bethe in his *Review of Modern Physics* article referred only to the paper by Bohr and Kalckar. Incidentally Bethe, in an earlier paper had referred to and dated this paper as being published in 1939 since he had seen the completed paper in 1936 and knew that Bohr would procrastinate.

Otto Robert Frisch (1904 – 1979) and George Placzek (1905 – 1955) set out to verify Bohr's ideas of the compound nucleus. To do this they needed to scatter neutrons off different materials. In particular they needed fairly thick targets of gold. For these they used several of the gold Nobel Prize medals that had been left with Bohr for safe keeping, after the Nazis came to power in Germany. They found a very large resonance in gold at a very low energy of only a few electron volts. This verified Bohr's theory beautifully. When Germany occupied Denmark, the gold medals were again in danger so they were dissolved in aqua regia where they remained in a jar on a shelf in full view until after the war. The gold was then recovered and the medals restruck.

It is not clear where Frisch developed his experimental technique, but he may have learned some tricks from his colleague Estermann. According to him, Immanuel Estermann claimed that there were two experiments that no physicist could do too often. One was to try to pump air through a stopcock that was shut. The other was to try to evacuate a vessel that had a stopcock to the outer atmosphere left open.

On a train trip from Berlin to Copenhagen Otto Frisch happened to be sharing the compartment with a handsome, dark skinned young man. After Frisch pulled out a novel by Edgar Wallace, the stranger asked, "You are a physicist?" The surprised Frisch asked him how he knew, to which the reply was, "You read Edgar Wallace." This is an example of how accurate a good physicists intuition can be. The stranger was Homi J. Bhabha (1909 – 1966).

Bhabha had come from Cambridge to Zürich to work with Pauli. He had a recommendation from R. H. Fowler who realized that Bhabha was very talented, but whom he also considered to be opinionated and arrogant. Thus he wrote to Pauli, "You can be as brutal a you like." Pauli was very pleased with that, "I can be as brutal as I like." As a consequence Pauli was very well-disposed to Bhabha and they became close friends.

From a rich Parsi family, Bhabha was also very handsome, well read, and a gifted painter, so much so that Casimir's wife liked to refer to him as the "fairy-tale prince". Later he went on to found the Tata Institute of Fundamental Research (TIFR) in Bombay. Shortly after he was killed in a plane crash on Mont Blanc.

Bohr's compound nucleus also gave an explanation for nuclear fission. The nucleus as a whole was treated as a 'liquid drop' and the incoming neutron got the drop oscillating so that it elongated, necked down, and finally broke apart into two almost equal drops. Lise Meitner and Otto Robert Frisch performed the calculation for thisprocess in the following sequence of events.

Lise Meitner (1878 – 1968) was an Austrian physicist who worked with Otto Hahn (1879 – 1968) for more than a quarter century. They verified some of Fermi's original discovery that by bombarding uranium with neutrons it was possible to produce atoms heavier than uranium. However, the results were ambiguous since there seemed a large number of other products. These were some of the products Fermi's group had thought to be transuranic. What made the situation even more complex, was that at the Paris Laboratory, Frédéric Joliot and Irène Curie found that there were products that were lighter than uranium.

After the *Anschluss* (the annexation of Austria by the Nazis) Meitner was suddenly a German citizen and subject to the new Nazi laws. Because she was Jewish she was afraid that she would not only lose her job, but also not be allowed to leave. So, with the help of some friends she secretly left her native country. Just before she fled to Sweden where she was offered a job at the Nobel Institute in Stockholm, Hahn had verified the Paris results. Some of the material behaved like radium, which was lower in the periodic table than uranium whereas adding a neutron with subsequent beta decay should have produced an element higher up in the periodic table. This result was very puzzling.

Once safely in Sweden Meitner received a letter from Hahn in which he informed her that the lighter material was not radium but instead was barium. This was even more surprising since nothing heavier than alpha particles had ever been observed to be knocked off a nucleus. She showed the letter to her nephew Otto Robert Frisch (1904 – 1979) who was visiting her for Christmas vacation. He thought there must be some mistake, but she assured him that Hahn was too good a chemist to make such a mistake. So, how could a nucleus that was like a liquid drop be split in half?

According to Frisch, they had been carrying on this discussion while walking through the woods in the snow. Now, he and his aunt sat down on a tree trunk and started back of the envelope calculations. The very large nuclear charge of uranium was indeed large enough to make the nucleus unstable. In fact the liquid drop model could be used to estimate the elongation and eventual splitting of the nucleus.

Although an excellent physicist, Meitner was not very musical. However, she learned to read music and play the piano. At one point she translated *"Allegro ma non tanto"* as "Fast but not auntie". The word *Tante* in German means aunt. She explained why she chose physics over mathematics as follows. "Mathematics began to seem too much like puzzle solving. Physics is puzzle solving, too, but of puzzles created by nature, not by the mind of man." This is not so different from a statement by Mark Kac. "If mathematics and science is looked upon as a game, then one might say that in mathematics you compete against yourself or other mathematicians; in physics your adversary is nature and the stakes are much higher."

After returning to Copenhagen, Frisch reported the results of his and Meitner's computations to Niels Bohr who was about to leave for the USA. Bohr immediately recognized the significance of this and asked if they had written up the result. When told that they had not, but were preparing to do so, Bohr promised not to talk about it until their paper appeared. Also Frisch wanted to perform some fairly straightforward experiments to verify all this. While writing the paper Frisch asked an American biologist what they called the process of cell division. That is how the word "fission" came to be the name for the process of nuclear splitting.

On the boat trip across, Bohr analyzed the ideas of Meitner and Frisch from all angles to convince himself. In this he was assisted by Léon Rosenfeld. On their arrival in New York, Bohr had some business to attend to and Rosenfeld went to Princeton accompanied by John A. Wheeler (1911 –) who had been waiting to meet them at the boat. In Princeton, Rosenfeld was invited to the "Journal Club" where recent results were discussed. He also was asked if he had anything new to report. Unaware of Bohr's promise and believing that the results of Meitner and Frisch were about to be published he let the story of fission out of the bag. Not only did this cause great annoyance to Bohr, but several Americans immediately set up experiments to verify the results with new experiments, experiments actually already performed by Frisch in Copenhagen and underway to be published. Some of the American results, however, were immediately reported to the newspapers. Even the *New York Times* of January 29 carried an article with the headline "ATOM EXPLOSION FREES 200,000,000 VOLTS; NEW PHYSICS PHENOMENON CREDITED TO HAHN." It was quite an effort to get the true priorities of the discoveries established.

This also led to an amusing misunderstanding. By now Bohr had heard about Frisch's experiments not from Frisch but from his son Hans. So, when a journalist asked Bohr how he came to know about these results he naturally answered, "My son wrote to me." The journalist's reaction must have been "His son. No, the name is Frisch. He must mean his son-in-law." Thus, it came about that Niels Bohr, who had no daughters, acquired, in the press, O. R. Frisch, a bachelor, as a son-in-law. This story was reprinted many times and shows how easily errors can propagate.

Bohr, because of the success of the compound nucleus, also strongly discouraged any investigation of the shell structure in the nucleus. This seriously delayed the understanding of the energy levels inside the nucleus.

The shell model, although delayed by the success of Bohr's theory, had been anticipated for a while. The reason was that for some time it had been noticed that nuclei containing certain numbers of protons or neutrons were particularly stable. The numbers for this stability are: 8, 16, 20, 28, 38, 50, 82, 126. Many theorists, who were still very much under the influence of Bohr's model, referred skeptically to these numbers as "magic numbers". The name has stuck and is no longer derogatory. The higher magic numbers: 50,

82, 126 gave rise to particularly great stability. Doubly magic nuclei, that is nuclei with a magic number of both protons and neutrons, are extraordinarily stable.

For example, tin with 50 protons has more isotopes (ten) than any other element. This suggested that just as in atoms where we get closed shells forming the very stable noble gases, so in nuclei there should be a shell structure to explain the stability associated with magic numbers. The key was to find a potential between the nucleons that would yield these magic numbers. Finally in 1949 Otto Haxel (1909 – 1998), Johannes Hans Daniel Jensen (1907 – 1973), and Hans E. Suess (1909 – 1993) in Germany and Maria Goeppert-Mayer (1906 – 1972) in the USA found the answer. The energy of the nucleons in the shell depends on their spin. They preferentially spin in the same direction as their orbit. That is, the energy is lower if the nucleon spins in the same direction as the orbit. So, in a new shell each proton (or equivalently each neutron) takes a separate orbit and thus all protons spin in the same direction as their orbit. When the orbit is half full, the proton can now either pair up with one of the protons in this orbit, but because of the Pauli exclusion principle it has to spin in the opposite direction. In this case it may take less energy to go to the next higher orbit. This explains why the magic numbers don't follow the same sort of regular pattern as in atoms.

The next stage in the progress of nuclear physics was due to the development of accelerators. The first accelerator or cyclotron was built by James Chadwick. However, one of the great promoters of cyclotrons was Ernest Orlando Lawrence (1901 – 1958) after whom the Lawrence-Livermore Laboratory is named. He began working with J. Robert Oppenheimer. In 1929 he invented the cyclotron, a device for accelerating nuclear particles to very high velocities. Hundreds of radioactive isotopes of the known elements were discovered using this device. Accordingly, element 103 was named Lawrencium (Lr) in his honor. For the development of the cyclotron, Lawrence won the Nobel Prize in Physics in 1939.

Lawrence was always very optimistic. In the same issue of the New York Times in which Rutherford's statement about nuclear energy being "moonshine" appeared, Lawrence was quoted, "I have no opinion as to whether it can ever be done, but we're going to keep on trying to do it."

On another occasion, he had meticulously modified pole pieces installed on the D-magnets of the cyclotron to minimize the fringing of the magnetic field. This had the negative effect of losing the beam. His reaction was, "Excellent! If it spoils the beam we will try the opposite effect and it will improve the beam." He turned out to be right.

In 1937, shortly after his brother, John Lawrence had shown that neutrons destroyed a deadly sarcoma in mice four times more effectively than X-rays, their mother was diagnosed with an inoperable cancer. The two brothers then subjected her to neutron radiation and she was cured.

There are two stories, about Lawrence, by anonymous physicists. The first dates from about 1938 in Berkeley. "The library wanted to put up an oil painting of Lawrence where everybody could see it. I heard that Segré had advised against it, so I told him it seemed a nice idea. Segré replied, 'You don't understand. Do you know anything at all about paintings?' He took me up to the library to look at it. It was horrible. I don't see how anybody could help seeing that it was horrible. Lawrence, however, was vastly pleased."

The other story has to do with taste buds. "A matter of taste buds. After the war I was with Lawrence and some of his crowd in a hotel room. They got out a bottle and offered me a drink. Rot gut! Terrible, terrible rot gut! Their faces fell apologetically when they watched me put it down. 'Something we've learned from recent researches at Berkeley', Lawrence explained. 'They've studied the aging process of whiskey there and identified the chemical compounds that stimulate the taste buds on the tongue. This bottle contains all the compounds and it would be a shame to waste money on a really expensive brand that doesn't contain any more.' " The comment finishes with, "These illustrate the 'accent on youth' that had turned merely to 'harshness of the palate.' "

At a seminar in Princeton, Ruby Sherr presented nuclear experimental scattering data which, he had been able to fit to a theory with four parameters. Since the fit was very good he proudly turned to Wigner who was, as usual, sitting in he first row and asked, "How many parameters should a good theory have?" Wigner's response was, "None."

In 1959 Wigner rhapsodized, "The miracle of the appropriateness of the language of mathematics for the formulation of the laws of physics is a wonderful gift which we neither understand nor deserve."

It may have been the atom bomb or the more practical aspects of nuclear physics which caused nuclear physicists to be better organized than other physics specialists. This did not always endear them to the rest of the physics community. Also the fact that early on, many nuclear physics experiments had to be rather rough led to the following statement. "In nuclear physics one equals ten for large one and small ten."

Chapter 6

Loss of Innocence: The Atom Bomb

"We couldn't have done anything else, but we have killed a beautiful subject."
Mark Oliphant.

At zero plus one minute after the first atomic bomb was tested, Kenneth Tompkins Bainbridge (1904 – 1996) turned to J. Robert Oppenheimer (1904 – 1967) and said, "Well, we're all sons of bitches now." Oppenheimer who was also an expert in Sanskrit and had read the *Upanishads* as well as the *Bhagavad Gita* in the original, quoted from the *Gita*, "I am become death, the destroyer of worlds." Two months after Nagasaki (October 16, 1945) he further declared, "If atomic bombs are to be added to the arsenals of a warring world or to the arsenals of nations preparing for war, then the time will come when mankind will curse the names of Los Alamos and Hiroshima." Again, two years later, in an interview with *Time Magazine*, February, 1948, Oppenheimer stated, "In some sort of crude sense which no vulgarity, no humor, no overstatement can quite extinguish, the physicists have known sin; and this is a knowledge which they cannot lose." When asked why he and other physicists would then have worked on such a terrible weapon, he confessed that it was "too sweet a problem to pass up".

Hans Bethe in an interview recalled his reaction to the bomb test. "The first reaction was, 'We've done it!' The second eaction was, 'What have we done!'"

The attempt to build an atom bomb in the USA began with a visit to the home of Albert Einstein by three physicists of Hungarian descent: Leo Szilard (1898 – 1964), Edward Teller (1908 – 2003), and Eugene Wigner. They were alarmed by the fact that Germany had passed a law forbidding the export of Uranium. The purpose of their visit was to persuade Einstein, who had good contacts with the Belgian royal family, to write to them to stop selling Uranium from their mines in the Belgian Congo to Germany. They also got him to add his signature to a letter to President Roosevelt urging the investigation of the possibility of constructing such an atom bomb. Szilard's main concern was that the Germans might be the first to develop such a weapon. As early as 1934, Szilard had realized that a chain reaction might be feasible in the right material and had even taken out a patent on such a process. At that time he was living in Britain and, for security reasons, had assigned the patent to the British Admiralty Office.

This was not the only patent that Szilard filed. He was a prolific inventor and even had several patents together with Einstein for a refrigerator without moving parts. He also invented various types of particle accelerators.

Szilard, later became interested in biology and published in this field. He then wrote to a physics friend, "There are only two kind of scientists among biologists: sons of bitches and bastards. The former are those who write papers and put in things that are false, the latter are those who point this out." He also advised Dennis Gabor (1900 – 1979), the inventor of holography, "There is no need to study mathematics. One can always ask the mathematician."

Szilard was also well known for his ability to tell Jewish jokes. One evening he and Victor Weisskopf engaged in such a joke telling competition. After about an hour Weisskopf excused himself to go to the washroom where he had previously hidden a book of Jewish jokes. However, after several such trips he had to nevertheless concede defeat when he realized that Szilard was making up the jokes as he went along.

To avoid the flood of published material Szilard suggested that when you got your Ph.D. you should be issued with 100 slips good for one publication each, no more. This way you would be much more selective about what you publish.

One other legacy from Szilard is the unit the "bit" of information. This is the smallest amount of information. He introduced this unit in a paper on the "Maxwell Demon" in 1929. Maxwell had conceived, of a way to violate the second law of thermodynamics that entropy must increase in any natural process, by imagining a demon that stood at an opening separating two parts of a container of gas. This demon would allow low energy molecules to pass from the right to the left, but reflect high energy molecules moving from right to left. In this way spontaneous cooling of the left side and spontaneous heating of the right side would occur.

When Szilard announced to Hans Bethe that he intended to keep a diary, "I don't intend to publish it; I am merely going to record the facts for the information of God," Bethe replied, "Don't you think God knows the facts?"

"Yes, but he does not know *this version* of the facts."

Szilard had not been the first to realize the feasibility of a nuclear chain reaction. The intuition of a great physicist is truly remarkable. In 1932, right after the neutron had been discovered, Rudolf Peierls (later Sir Rudolf) was travelling the Caucasus Mountains with Lev Landau. He casually asked Landau if he thought a nuclear explosion could be achieved. After thinking for a long time Landau replied, "I think it is quite possible if the neutron capture cross section of a nucleus turns out to be sufficiently large and we learn to direct nuclear reactions." This, as we know, proved indeed to be the case.

Rudolf Peierls was German born, but, since his wife was Jewish, he fled to Britain after the Nazis came to power. As a German born, he was not cleared for war research in Britain. At the University of Birmingham the adherence to secrecy was maintained as follows. At tea time Mark Oliphant, the head of the physics department would approach Peierls with an "hypothetical" problem. "Suppose you had to solve Maxwell's equations for a conducting cavity of the following shape, how would you proceed?" Both Peierls and Oliphant knew that this was a problem connected with the generation of radar waves, but both pretended that the problem was of purely academic interest. Peierls typically responded that the problem was very interesting and a few days later he would return with the solution.

Even stranger is the fact that although Peierls and Otto Frisch had been the first ones to calculate the critical mass required for an atom bomb Peierls, being a foreigner, was for security reasons not allowed to discuss his own work.

That Peierls had a sense of humour is evident from the following episode. As holder of a Rockefeller Foundation grant in Cambridge, Peierls like all such grant holders, had to submit an annual report listing his achievements during the period. As part of the proof of his activities Peierls sent the foundation an announcement of the birth of his first child. The officials at the foundation were not amused.

Incidentally, Mrs. Peierls defined an anti-Semite as "someone who hates the Jews more than is absolutely necessary."

The results obtained by Otto Hahn and interpreted by Lise Meitner and her nephew, Otto Robert Frisch, as discussed in the previous chapter, indicated that a self-sustaining nuclear reaction or even explosion might be possible. To verify this would require constructing a nuclear reactor or "pile", as it was then known.

After getting Einstein's signature, Szilard, Teller and Wigner tried to get the funds to attempt to construct a nuclear pile to show that a chain reaction was indeed possible. For this they needed about $6, 000.00 to buy some very pure Carbon and Uranium oxide. This was a rather substantial amount of money at the time. The US ordnance expert, a Colonel Adamson, who reviewed their grant application considered with skepticism these physicists trying to describe a possible new type of bomb. Finally he declared that to develop a new weapon normally requires two wars and that anyway the most important thing in war is morale. To this Wigner, in his quiet way, replied, "Perhaps it would be better if we did away with the War Department and spread the military budget among the civilian population. That would raise morale a lot." They got the $6, 000.00.

After Szilard, Teller and Wigner secured the funding, Fermi constructed the first nuclear pile (reactor) in a squash court, under the stadium of the University of Chicago. It was called a "pile" because it consisted mainly of a huge pile of very pure carbon bricks interlaced with Uranium oxide and rods of Cadmium. Carbon slows the neutrons down so that they can be captured by the fissile Uranium 235 whereas Cadmium strongly absorbs neutrons and so can be used to control the reaction. As the Cadmium rods were slowly and carefully pulled out of the pile, the counters started clicking more and more rapidly. Finally, enough rods had been removed and the count leveled off. The first self-sustaining nuclear reactor had become operational. Eugene Wigner pulled a bottle of Chianti from behind his back and presented it to Fermi. Everybody present drank a glass.

That evening, Fermi and some colleagues celebrated in Fermi's home with champagne, but his wife was not informed what the celebration was about and did not find out until after the war. Compton also reported from Chicago to Conant by telephone in code that the first chain reaction had been successfully initiated by Fermi, "The Italian navigator has reached the New World. And how did he find the natives? Very friendly." This was the crucial experiment that showed that it might be possible not only to extract energy from the nucleus, but also to construct a bomb.

The empty Chianti bottle also has a history. It was gathered by Al Wattemberg, signed by all present and saved by him. He promised to bring the bottle to Chicago on December 2, 1952 to celebrate the tenth anniversary. However, on that day his wife gave birth and so he could not attend. Instead he mailed the bottle, but to make sure that it would arrive safely he insured it for $1, 000.00. This was a very large amount of money in those days and newspapers spread the story. As a consequence an importer of wine sent Fermi a whole case of Chianti to thank him for all the free advertising.

Fermi, whose wife was Jewish, had escaped from Mussolini's Italy, after he received the Nobel Prize. Bohr had forewarned him that he was to receive the Prize and so, on a visit to Stockholm, he had taken his family with him. From there they proceeded directly to New York where he had a job waiting for him at Columbia University. Once settled, he tried to become Americanized as quickly as possible. At this time, the Fermis lived near the home of Harold Urey (1893 – 1981) who had a Nobel Prize in chemistry, for isolating deuterium. In a seminar at Columbia University, Urey gave a cost estimate for the production of 99% pure heavy water by the fractionation method and included the following statement, "assuming that the graduate student's time was worth nothing."

Urey volunteered his help and instructed the Fermis in the mysteries of this most American of activities, the maintenance of a lawn. After he had inspected the Fermis' lawn he declared it to be nothing but crab grass and that this was a weed. Now, Fermi liked to classify things in order to be able to better understand their relationship to other things. So, he asked what was wrong with crab grass and why he should be concerned. After all, it was green and covered the lawn. What was it that distinguished a weed from any other plant? To this Urey responded that weeds grow without having been planted and take up the nutrient that good plants require. At the end of a season they die and there is nothing to show for them. At this point Fermi had his classification, "Weeds are unlicensed annuals."

In a sense, the construction of the atom bomb was inevitable once the possibility of such a destructive weapon was realized. This enormous enterprise also brought together, on a remote mesa in New Mexico, a collection of some of the world's greatest physicists. In June 1942, J. Robert Oppenheimer (Oppy to his friends) was appointed the technical Director of the Manhattan Project. Under his guidance, the laboratories at Los Alamos were constructed. There he brought some of the best minds in physics to work on the problem of creating an atomic bomb. Within a short time this became the biggest physics research center in the world. Of all the great physicists, involved in the study of the nucleus or high-energy physics, in the USA at the time one that did not join this enterprise was Julian Schwinger, who worked on radar. In the end Oppenheimer was managing more than three thousand people, as well as tackling theoretical and mechanical problems that arose. He is often referred to as the "father" of the atomic bomb.

After sitting in on one of his first conferences at Los Alamos, Fermi turned to Oppenheimer and in a surprised voice said, "I believe you people actually want to make a bomb." He also stated, "Whatever Nature has in store for mankind, unpleasant as it may be, men must accept, for ignorance is never better than knowledge." On the other hand, with respect to the Super, the hydrogen bomb, he had this to say, "The fact that no limit exists to the destructiveness of this weapon makes its very existence and the knowledge of its construction a danger to humanity as a whole; it is necessarily an evil thing, considered in any light." While working on the Super Bomb, Fermi told Ulam, "I'm not sure the human race is going to survive."

Like other famous physicists, involved in nuclear research during the war, Fermi was given an alias, Mr. Farmer, as well as a bodyguard. Also like many physicists, Fermi loved to go for long walks and was accompanied on these by his bodyguard. To break the silence Fermi talked physics to his guard. After several weeks of this he confided to another physicist, in his strong Italian accent, "My bodyguard, he learn so much physics, soon he himself need a bodyguard."

Incidentally, during all this time Fermi was still officially an enemy alien. He and his family did not become US citizens until 1944. There is a very revealing story of Fermi's ability to use the simplest experiments to obtain a result. When the test of the first atomic bomb was scheduled at the appropriately named Jornada del Muerto (Journey of Death), near Alamogordo, New Mexico Fermi, as an alien, was not allowed to be present, even though he had been instrumental in its development. So, at the time of the explosion he sat in front of his hut some twenty kilometers away and slowly dropped bits of paper from chest height to the ground. When the shock wave from the blast passed, his pile of papers moved some distance, which he measured. He then used this to calculate the force of the blast. Later, when his colleagues returned from the blast site with their measurements he informed them of the amount of energy released in the explosion. They questioned his number, since they had not yet analyzed their data. After the analysis it turned out that his estimate was within 10%.

It may have been his extraordinary abilities that lead Fermi to claim that any physicist worth his salt should be able to estimate, within a factor of three, any number in the universe that had something to do with an observable quantity. He was immediately challenged, "How many locomotives are there in the USA?" Another version has him estimating the number of piano tuners. After a short computation he came up with a number that, after some research, proved to be within the desired factor of three.

Another illustration of his extraordinary abilities may be seen from the following. On a train trip Fermi, A.H. Compton, and S. K. Allison were travelling over mountains and Fermi appeared bored. Compton, to give Fermi something to amuse himself, presented him with the following problem. While doing his cosmic ray experiments in the Andes he had noticed that his watch did not keep time as accurately as at sea level and he wondered if Fermi could explain this. Fermi perked up, his eyes twinkled and out came his pocket slide rule. For the next while he was busy scribbling equations of damping of the balance wheel due to the air and the effect of the lower pressure of the air at high altitude. Some five or ten minutes later he was finished and presented the result of his calculations. Compton was most amazed to see that the numbers agreed with his recollection of the behaviour of his watch.

While working on the theory of a nuclear reactor Wigner used an exact formulation that involved integro-differential equations. Fermi simply smiled, "Poor guy, see how he suffers solving them." Fermi used a physical approximation and obtained the desired result with sufficient accuracy in a straightforward manner.

Fermi was also a great admirer of von Neumann. When one of Fermi's close friends asked him about von Neumann he received the following response. "What do you think about me? Well, you should know that he is as more intelligent than I am, as I am more intelligent than you." Apparently this did not offend Fermi's friend since he was quite aware of Fermi's abilities.

Hans Bethe had this to say regarding Fermi, "Fermi was always willing to help with specific problems; anyone could come to him."

While at Los Alamos, Fermi took up fishing. However he refused to use artificial flies or lures because, "the condemned deserve an honest meal."

After the war, German and Japanese physicists were not permitted to do nuclear research. Italian physicists, although not so prevented, did not engage in nuclear physics because they did not think that they could compete with the Americans who had such a big head start. Also there was another factor, as explained by Edoardo Amaldi (1908 – 1989) after a meeting with Fermi in 1946. "At a certain moment I found, when we started talking about neutrons, he was talking completely freely up to a certain point, and then it was clear he did not want to give more information - not because he did not want to, but he could not talk about fission. I found that extremely unpleasant. So, I did not want to work in a field where people were not able to talk freely."

From the repeated exposure to radiation, while working on these bombs, Fermi developed cancer and in 1954 while in the last stages of this disease he railed against the laws prohibiting euthanasia, "They put dogs out of the way. Why must I go through this?"

Another of the famous physicists that arrived at Los Alamos was Niels Bohr. He had kept in touch with James Chadwick who was a sort of unofficial leader of the British atomic project. At the beginning of 1943 Chadwick had sent a secret message in a micro-dot hidden in a key handle, inviting Bohr to escape to Britain. At this time Bohr decided to stay in Copenhagen since he was still skeptical about the possibility of constructing an atom bomb. However, by September the situation had changed and Bohr was in danger of being arrested. So, he escaped.

For this escape, Bohr first visited neutral Sweden and from there sneaked out to England in a Mosquito bomber. Due to Sweden's neutrality, the plane was completely unarmed, but had been chosen because its great speed gave it the chance to cross the German territories safely. The plane could carry only a single passenger in the bomb bay at the rear of the plane. To avoid anti-aircraft fire, the plane had to fly at great height. During the trip the pilot tried to communicate with Bohr to turn on the oxygen, but got no answer. He realized that something was wrong and dropped to lower altitude once he crossed Norway. Upon landing in Scotland, he hurried back to check and found Bohr fast asleep.

What had happened was that the headphones did not fit Bohr's rather large head and he never heard the pilot telling him to turn on the oxygen. Thus, he soon fainted. By the time they landed in Scotland, Bohr had revived and was indeed dozing, but refused to admit that he had fainted.

Bohr's arrival in the USA during the war was supposed to remain secret. Accordingly, he was met at the boat by two FBI agents who grabbed his and his son's suitcases and hustled them into a taxi and from there through the reception into their hotel room. Finally, after this rush, they allowed them to relax, secure that Bohr's identity was safely hidden. However, at this point one of the agents noticed that on Bohr's suitcase in bold black letters was written NIELS BOHR.

There are numerous stories about Bohr's security problems. During the war on his visit to the USA Bohr was, for security reasons, given the alias "Nicholas Baker". He was affectionately called, "Uncle Nick". When asked whether he found it difficult to remember to sign himself that way rather than by his real name, he responded, "What's the difference? My signature is so illegible it could be either."

On a trip to Washington Bohr encountered in an elevator a young woman he had known in Copenhagen as the wife of the nuclear physicist, von Halban. She immediately recognized Bohr and greeted him, "So glad to see you again, Professor Bohr." Anxious not to break security, yet remain polite, Bohr replied, "You must be mistaken, my name is Nicholas Baker. But, I do remember you. You are Mrs. von Halban."

"No, I am Mrs. Placzek." came the short reply. She had been divorced and remarried another physicist, George Placzek.

While shopping for some clothes, Bohr had left his watch in the store. He had given his name as "Baker", but the watch was engraved "Bohr". His son, Aage Bohr, went to retrieve the watch and pretended that he was James Baker, secretary to Mr. Bohr. He pointed out that the store could confirm this by phoning Mr. Bohr at the Danish Legation. The store phoned and Bohr was out, but a member of the Legation staff informed the caller that Mr. Bohr's secretary was also called Bohr. At this point Aage Bohr came clean.

Although Bohr had excellent command of the English language there were some mispronunciations that were an integral part of his speech and are worth recording. He always referred to the atomic bomb as the "atomic bum". More interesting, perhaps as a Freudian slip, is the fact that he always referred to the FBI as the FIB.

Among all his other accomplishments, Bohr was an excellent skier, so much so that once on a skiing tour to Dalarne one of the Swedish students was heard to make the following complimentary remark, "The only criterion that the professor is a professor is that the professor always forgets his gloves." This would have been useful knowledge for a young American physicist who did not know Bohr. While Bohr was in the USA "incognito", working on the atomic bomb project, there were occasional ski outings. A younger physicist, who did not know him, saw this middle-aged man without skis staring wistfully at the slopes. Thinking that such a man would certainly not be able to ski for more than a couple of hours he offered to lend him his skis. Bohr accepted gratefully and returned them after it got too dark to continue skiing.

The name of the group working on the nuclear bomb in Britain during World War Two was the Maud Committee. How this name came to be chosen is as follows. Lise Meitner co-authored with Otto Robert Frisch the paper that gave the theoretical explanation of the Uranium fission discovered by Hahn and Strassmann. In May 1940, just after Germany invaded Denmark, Meitner sent the following telegram from Sweden to the British physicist O. W. R. Richardson.

> Met Niels and Margarethe recently.
> Both well, but unhappy about events.
> Please inform Cockroft and Maud Ray Kent.

Cockroft thought that the last three words were an anagram for "radium taken" and so these were deemed a good code name. The mystery about these last three words was cleared up when, in 1943, Bohr arrived in England and asked whether the message had reached his old governess Maud Ray who lived in Kent.

Edward Teller (1908 – 2003) was also one of the famous people involved with Los Alamos. He is the father of the hydrogen bomb. A very proud man, he was not content to play second fiddle and attempted to challenge Oppenheimer's leadership at the Tuesday meetings at Los Alamos. One time during the discussions of a thermonuclear device, Teller got up to make a speech. He started out by saying that he would give only qualitative factors. But when he warmed up he laid out a few calculations which showed he had actually forgotten a factor c^2, the speed of light squared. Apparently, Oppenheimer stared at the blackboard as though it opened totally new concepts. He must surely have wondered what sort of reasoning could operate with an error of a hundred trillion per cent. Oppenheimer asked, "This idea of dealing only in qualitative factors makes an interesting approach, but should we go so far as to treat the speed of light as unity?" This story may also explain why Teller gave such damning testimony against Oppenheimer at the McCarthy Senate Investigation Committee. "I feel I would prefer to see the vital interests of this country in hands that I understand better and therefore trust more."

While working on the Super Bomb, Teller was a real slave driver. When Frederic de Hoffman approached Teller to tell him, "I'm getting married today," the latter replied, "Freddy, does this mean that you will not be at work in the lab tonight?"

Here are a couple of quotes from Teller. "A fact is a simple statement that everyone believes. It is innocent, unless found guilty. A hypothesis is a novel suggestion that no one wants to believe. It is guilty, until found effective." Also, "Two paradoxes are better than one; they may even suggest a solution."

During the Manhattan project the mathematician, Stanislaw M. Ulam (1909 – 1986) worked with von Neumann in the Theoretical Division at Los Alamos. Here he developed the Monte-Carlo method for solving complex problems. This method is still very much in use by high energy experimental physicists. Later he checked whether Teller's design for a hydrogen bomb would work. Ulam and Cornelius Everett, also a mathematician, concluded that Teller's model would never work. This caused quite some friction between Ulam and Teller. However, after about a year, Ulam came up with a design that made the hydrogen bomb possible.

In an autobiography, *Adventures of a Mathematician*, Ulam wrote what should be a warning to all potential chairs of academic departments. "When I became chairman of the mathematics department of the University of Colorado I noticed that the difficulty of administering N people was not really proportional to N but to N^2. This became my first 'administrative theorem'. With 60 professors there are roughly eighteen hundred pairs of professors. Out of that many pairs it was not surprising that there were some that did not like one another."

Isidore I. Rabi (1898 – 1988) was another eminent physicist at Los Alamos. He had studied molecular beam techniques with Otto Stern and later used these techniques to measure the magnetic moment of the electron. This was an important step towards the development of quantum electrodynamics. Later he used the technique to develop what became known as nuclear magnetic resonance in the field of medicine but has now been renamed as MRI or magnetic resonance imaging so as to avoid the word "nuclear".

At one point Rabi was trying to explain the magnetron (a 10 cm microwave cavity) to a group of physicists. "It's simple; it's just a kind of whistle." To this Edward Uhler Condon replied, "Okay Rabi, how does a whistle work?" Rabi could not answer.

Here is an example of a warning to high energy physicists issued by Rabi much later in his life. At the Niels Bohr symposium October 3, 1985 Rabi cautioned, "Often new discoveries are made by closer examination or an increase in accuracy of measurements of already known results. This is not possible any more. Often I listen to talks in experimental high-energy physics and I ask the speaker, 'have you published this?' He answers, 'Of course I have', but really he has not. Only conclusions are published, not experimental data that can be subjected to criticism. Moreover, the experiments are too expensive to be repeated. The experimental physicists are reduced to technicians testing some theoretical predictions. I am afraid that what we see today sometimes may be artifacts of the theory."

On viewing the paintings of Hans Memling in the museum of the same name in Bruges, Belgium, Rabi mused, "When an artist is good, he can be very great." He also advocated eliminating three-tined forks from tableware in favour of four-tined forks since the former were a "totally impractical tool".

At Los Alamos, Rabi became known as the Rainmaker. The story that follows is as he related it. Wherever Rabi went he wore rubbers and carried an umbrella. So when he showed up at the first atomic test site in the baking noon of mid-July on the sands of New Mexico he stepped out with the inevitable rubbers and umbrella. After he had placed his dollar in the betting pool on the strength of the bomb he received Oppenheimer's instructions. "You will room with me. Change to desert clothes. One reason we picked this spot is that it won't rain here. Naturally I have checked with the meteorologists." No sooner had Rabi changed than a downpour reminiscent of a monsoon began.

Later, Rabi became an advisor to President Eisenhower. By this time he was a Nobelaureate, and so it was natural that he introduced his fellow Nobelaureate, Hideki Yukawa to the President. Eisenhower shook Yukawa's hand and said, "Finally I have shaken the hand of a Nobelaureate."

After the launching of Sputnik when Rabi again met President Eisenhower, whom he had known since Eisenhower's stint as president of Columbia University, the latter looked at Rabi and said, "You seem a bit annoyed with me." Rabi replied, "You seem to have ignored the scientific community up to this point. I trust you will make up for it."

Rabi also did not mince words. Regarding Lawrence's super-giant linear accelerator the MTA (Materials Testing Accelerator) which was for manufacture not testing he had this to say, "You can make anything defy the laws of physics, at least for a while, if you spend enough money on it, The MTA was silly."

After the hydrogen bomb was built, Oppenheimer responded to journalists:

"If you ask: can we make them terrible, the answer is yes.

"If you ask: can we make more of them, the answer is yes.

"If you ask: can we make them terribly more terrible, the answer is — probably."

It may be the development of the hydrogen bomb that prompted Oppenheimer to say, "The optimist thinks that this is the best of all possible worlds, and the pessimist knows it."

About the departmental tea, Oppenheimer had this to say. "Tea is where we explain to each other what we don't understand." Also, with his predilection for Sanskrit, it was no surprise that Oppy named his fiery-red sportscar "Garuda". This is the name of the god Vishnu's vehicle.

Perhaps the most entertaining and brilliant character in the Los Alamos story was Richard Philips Feynman (1918 – 1988). Apparently, during the Manhattan project there were frequent loud discussions between Feynman, "the mosquito boat" and Hans Bethe, "the battleship". Feynman was hear to shout objections or else exclaim admiration at almost every sentence uttered by Bethe. Bethe just continued to calmly plow ahead and calculate through these tirades. Hence, the names given by their colleagues. Bethe was also the division head. This was a position Teller thought that he should have and was another reason for him to dislike Oppenheimer.

While working at Los Alamos on the atom bomb, physicists and their families were confined to a fenced compound for security reasons. Of course there was a hole in the fence which everyone, except the security personnel, knew about. Feynman used this hole to drive one of the guards to distraction. He crawled out through the hole and entered through one of the main gates. Then he repeated this time after time. The poor guard must certainly have wondered how a man could reach the other gate, more than a mile away, in such a short time.

There was also no safe that was secure if Feynman wanted to get into it. He had realized that physicists would be prone to use certain numbers and he would try various combinations with success. This also led to the following story when someone asked him if physicists were getting anywhere with answering the "big question". He responded, "You ask, are we getting anywhere? I'm reminded of a situation when I was asked the same question. I was trying to pick a safe. Somebody asked me, 'How are you doing? Are you getting anywhere?' You can't tell until you open it. But you have tried a lot of numbers that you know don't work!"

Feynman also often demonstrated his ability on the bongo drums. One time, he had one half of his physics class count the number of beats with his left hand and the other half count the beats with his right hand. To the class's amazement he was able to do twelve on seven.

The following story about Richard Feynman may or may not be true since I have been unable to verify it. When Feynman was sent the letter from Uncle Sam that began with "Greetings" to report for his physical for the draft into the military, he did so promptly. However, the tedious lineups for the various examinations took most of the day and he got bored. So, when his final test with a psychologist for mental fitness came up, he decided to have some fun. As he walked into the office he kept looking over his shoulder as if he someone were following him. When the psychologist asked him to hold out his hands he did so with one hand palm up and the other palm down. When he was asked to turn his hands over, he turned both so that again one palm was up and one down. These antics continued throughout the interview. Finally he was dismissed.

A few weeks later he received his classification as 4F, in other words unfit to serve. The reason was "mentally unfit". After Feynman stopped laughing, he felt guilty and wrote to his draft board that he was really quite mentally fit and had only been putting on an act. The response was, "They all say that."

On one occasion, millions of people were able to see Feynman's genius first hand. Part of the investigation into the 1986 disaster of the space shuttle Challenger was televised. Feynman was on the commission. To demonstrate the cause of the disaster he plunged an O-ring, used as a seal, into a bucket of ice water. Later he pulled it out and showed that it had become stiff and thus due to the unusually cold weather in Florida at the time of the launch, had failed to provide a proper seal, causing a fuel leak that led to the disaster. At the time he also estimated that for such very complicated machinery the failure rate was closer to 1 in 100 than the 1 in 100, 000 claimed by NASA.

When Feynman was on his deathbed in a coma, his wife Gweneth was beside him. It seemed to her that he moved his hand as if he wanted her to hold his. She asked the doctor if she was right. The doctor told her, "It's purely automatic; it doesn't mean anything." This was when Feynman, who had been in a coma for 36 hours, made his final declaration about experts. He lifted his hands, shook out his sleeves and folded his hands behind his head.

Another physicist involved in the Manhattan project was James Franck one of the two collaborators of the Franck-Hertz experiment. During World

War One, he had joined the German army and was immediately given command of a platoon. On the first day he shouted, "Platoon, turn right, please!" His promotion thereafter was considerably retarded. Later, in 1935, he emigrated to the USA where he first became a professor at John Hopkins University and later at the University of Chicago and University of California.

Even though he worked on the atomic bomb project, he later also headed a group of Chicago scientists that urged the American government not to use the atom bomb against a Japanese city, but to demonstrate its destructive power to the Japanese so that they could surrender. This was the *Franck Report* submitted to the Secretary of War. In this report he predicted, "If the United States were to be the first to release this new means of indiscriminate means of destruction of mankind, she would sacrifice public support throughout the world, precipitate the race for armaments, and prejudice the possibility of reaching an international agreement on the future control of such weapons."

After the war, his erstwhile friend and collaborator, Robert Wichard Pohl (1884 – 1976), was due to arrive as a visitor at the University of Illinois. Before the war they had been colleagues at Göttingen, but had a falling out over campus politics. Furthermore during the war they were on opposite sides. Prior to Pohl's arrival, Franck telephoned Fredrick Seitz at Illinois to ask him to tell Pohl that under no circumstances would he be welcome at the University of Chicago where Franck was now located.

A few weeks later Pohl arrived at Illinois and after his short stay stated that his good friend, Franck, was now at Chicago and since that this was so near he was certainly going to stop and see him. Seitz tried hard to dissuade him. To no avail. Finally he even revealed the details of Franck's phone call. Pohl simply "pooh-poohed" the whole notion of enmity and departed for Chicago. A week later Pohl returned from Chicago. His return was followed by a phone call from Franck thanking Seitz for "insisting" that Pohl visit him. Apparently, as soon as the two old friends saw each other again, all bitterness disappeared and they fell into each other's arms.

When Victor Weisskopf (1908 – 2002) worked on the atomic bomb project at Los Alamos, he was constantly consulted by experimentalists about the size of various effects, so that they would know how detailed an experiment was required. After a while someone put up a sign in the hallway, pointing to his office, "Los Alamos Oracle".

Weisskopf had received his training in Europe. He had this to say about it. "In Munich and Göttingen you learned to calculate. ... In Copenhagen you learned to think. I must say that in Zurich with Pauli I learned both."

After Weisskopf arrived in Copenhagen for the first half of his Rockefeller fellowship he was met by his friend Max Delbrück (1906 – 1981) who told him that Copenhagen was wonderful with many beautiful girls. Weisskopf replied, "That is very interesting information, but has no value for me since I want to do physics." Delbrück countered, "Well yes, but at least you could have a look." So they went dancing and there, Weisskopf met the woman that became his wife.

Later Delbrück became a biologist and when Crick and Watson published the double helix he wrote to them. "I have a feeling that if your structure is true ... then all hell will break loose, and theoretical biology will enter a most tumultuous phase." In 1969 Delbrück received the Nobel Prize in medicine.

While Weisskopf was with Pauli, this gentleman advised him, "Don't become an expert, because of two reasons: First, you become a virtuoso of formalism and forget about real nature, and second, if you become an expert, you risk that you are not working for anything really interesting anymore."

Weisskopf published his first paper, on the width of spectral lines, in collaboration with Eugene Wigner. He considered himself very lucky that in this, his first paper, he was first author since although starting with a W, alphabetically his name preceded Wigner's. He vowed that henceforth he would maintain strict alphabetical order when co-authoring. This led to some difficulties when he co-authored the book on nuclear physics with John Markus Blatt (1921 – 1990). The publisher wanted Weisskopf, as senior author, to be listed first. Weisskopf was adamant that it should be Blatt and Weisskopf. He finally convinced the publisher by saying if it were the other way around the emphasis would be on Blatt.

When both were already fairly advanced in age, Victor Weisskopf met Otto Frisch at a conference. He listened to some of the talks and turned to Frisch, "Those young men are talking a lot of nonsense, it's up to us more senior ones with knowledge and courage to tell them where to get off." Frisch was more hesitant, "But I'm not sure how to tell if what they are saying is nonsense." Weisskopf was more certain, "That's where courage comes in."

On another occasion he lamented, "We have bred too many experts and there remain too few scientists."

At the beginning of a speech to the Royal Society of Canada, Gerhard Herzberg, a physicist, but with a Nobel Prize in Chemistry, attributed the quote, "If you understand hydrogen, you understand all that can be understood," to Weisskopf. Hydrogen and its isotopes were close to Herzberg's heart since much of his research dealt not only with that atom, but also with the molecule. When Herzberg wrote to Weisskopf to ask him if this attribution was correct, the latter replied that he did not remember making the statement, but did not mind having it credited to him since it was correct.

Later Weisskopf (Viki) became the director general of CERN, [1] a position he held during its very productive period from 1961 to 1965. In 1964 he presented a talk at CERN with the title, "The Significance of Science". During this talk he said, "Human existence is based upon two pillars: compassion and knowledge. Compassion without knowledge is ineffective; knowledge without compassion is inhuman."

The German atom bomb project never even achieved a nuclear reactor. This was very surprising since Otto Hahn had been the first to find nuclear fission. The reasons for this are several, but some of the main ones are that the German government never funded this project to an extent necessary. The German effort was also spread among several competing professors and was never as coordinated as the American effort. Another reason is that early measurements by Walther Bothe indicated that carbon was not suitable as a moderator for nuclear reactions. This result was probably due to the carbon that he used being not sufficiently pure. The consequence of Bothe's measurement was that the German effort concentrated on using heavy water, most of which was obtained from a plant in Norway. The effort was further hampered when the British with he help of the Norwegian underground destroyed the Norwegian heavy water plant.

---

[1]CERN stands for Conseil Européen pour la Recherche Nucléaire and was originally proposed by Louis de Broglie as a European science laboratory. In 1950 Isidore Rabi proposed to UNESCO creation of regional centres for collaboration of scientists. This eventually lead to the birth of the world's largest research centre in 1954.

Chapter 7

Elementary Particles

"Nevertheless, it sometimes happens that we falsely believe that we have in our minds the ideas of things because we falsely suppose that we have already explained certain terms which we are using." Gottfried Wilhelm von Leibnitz, Reflections on Knowledge, Truth, and Ideas. (1684).

"If I could remember the names of all these particles I'd be a botanist." Albert Einstein.

At a symposium in 1972 in Trieste, Italy to celebrate the seventieth birthday of Dirac, I heard the following conversation between three of the greats of physics.

Dirac: "Is the electron elementary?"

Heisenberg: "No!"

Wigner: "It depends on the point of view. That is, the electron transforms according to an irreducible representation of the Poincaré Group."

Heisenberg: "Composite particles also transform that way."

Wigner: "It is a matter of definition. We can define a particle as elementary if its state is orthogonal to that of a composite particle."

The rest of the conversation became even more technical.

In response to Bohr's letter congratulating him on his on 60th birthday, Heisenberg had written, "It strikes me as almost strange to take part once again, as recently in Brussels, in the battle of opinions and carefully to weigh the various contradictory arguments, just as we did 30 years ago in Copenhagen. The young physicists watch this with some amazement, since they have probably become used to the notion that, in the end, if only sufficiently many physicists are placed at sufficiently big machines, then everything will fall into place in the end."

Takehiko Takabayasi also once asked, "What the devil is particle physics?" Sin-Ichiro Tomonaga had an even more general question. "What the devil is physics?"

In 2001 at the Heisenberg Centennial in Bamberg, someone asked the Nobelaureate, Martinus J. Veltman (1931 –), "Can you tell me what is an elementary particle?" His reply was, "I can't tell you what it is, but I can tell you that you can make lots of money with it." The reason for reporting these conversations is to show how complicated this whole business of elementary particle physics really is.

The first elementary particle to be discovered was the electron, soon to be followed by the proton. Then, in 1932, Chadwick discovered the neutron and Carl D. Anderson the positron. Anderson published his results in 1933 in a paper with the title *The Positive Electron*. There was also a somewhat earlier paper by Lord Patrick M. S. Blacket (1897 – 1974) and Giuseppe Beppo Occhialini (1907 – 1993), to which he refers. Although this verified the prediction by the Dirac equation that such a particle should exist, none of the authors referred to Dirac's work.

By the end of 1933 at the Seventh Solvay Congress, Lord Blackett was able to report that further work on gamma rays showed that the Dirac theory was correct. Rutherford, who was present, lamented, "It seems to a certain degree regrettable that we had a theory of the positive electron before the beginning of experiments." He continued, "I would be more pleased if the theory had appeared after the establishment of the experimental facts."

Pauli's neutrino, invented to explain beta decay, added to the proliferation of elementary particles. As if not enough, the neutrino was also involved in the decay of the pion into a muon and a neutrino only, as shown by Leon M. Lederman, Melvin Schwartz, and Jack Steinberger, this was another kind of neutrino. The trio shared the 1988 Nobel Prize in physics "for the neutrino beam method and the demonstration of the doublet structure of the leptons through the discovery of the muon neutrino." Here is a quote from Lederman regarding physics funding. "Physics is not a religion. If it were, we'd have a much easier time raising money."

As mentioned earlier, a big theoretical break-through to understanding the strong nuclear force came in 1935. In Japan Hideki Yukawa, only twenty-nine years old, proposed the existence of a new particle roughly 200 times as heavy as the electron and that this particle would jump back and forth between the constituents of the nucleus to produce the strong attractive force between them. In 1936 Carl Anderson together with Seth Neddermeyer found the muon. At the time they were unaware that Hideki Yukawa had predicted a particle with a mass about 200 times that of the electron. After Yukawa's work became known most physicists believed that the particle that Anderson and Neddermeyer had discovered was Yukawa's meson. This was not to be.

Hideki Yukawa was born Hideki Ogawa. His father-in-law adopted him and he changed his name to Yukawa when he got married. This was not unusual when a Japanese bride had no brothers. His father too had been adopted in this manner and changed his name from Asai to Ogawa.

In a lecture that he presented in 1974, Yukawa referred to the general attitude with regard to physics in the period around 1926 when he was about to enter Kyoto University.

"At this period the atomic nucleus was inconsistency itself, quite inexplicable. And why? — Because our concept of elementary particle was too narrow. There was no such word in Japanese and we used the English word — it meant proton and electron. From somewhere had come the divine message forbidding to think about any other particle. To think outside of these limits (except for the photon) was to be arrogant, not to fear the wrath of the gods. It was because the concept that matter continues forever had been a tradition since the time of Democritus and Epicurus. To think about the creation of particles other than photons was suspect, and there was a strong inhibition of such thoughts that was almost unconscious."

Not only did Yukawa predict that the mass of this as yet undetected particle had to be about 200 times that of the electron but also that the strength of the interaction of this particle with protons and neutrons had to be about 1000 times that of the electromagnetic force. This non-existent particle had, at first, many different names: yukon, japanese electron, mesotron, heavy meson, meson. As stated, success seemed to have been achieved when Anderson found a particle with a mass 207 times that of the electron. However, the interaction strength of this particle was about the same as that of an electron and thus could not possibly account for the strong nuclear force.

The puzzle was resolved in 1947 when Giuseppe Beppo Occhialini and Cecil F. Powell (1903 – 1969) found, in extremely high atmospheric cosmic rays, a very strongly interacting particle with a mass of about 264 times the electron mass. This was the real sought-for Yukawa particle. Today it is called the pion (pi meson) and there are three different pions: a positively charged, a negatively charged, and a neutral.

The cloud chamber had provided a wonderful tool for studying the tracks left by charged particles. However, cloud chambers are filled with gas and interactions occur only when particles collide. Since gases lack density thessse collisions are infrequent. To produce large numbers of collisions with atoms it was necessary to insert metal plates in the cloud chamber. These collisions occurred inside the metal plates and thus could not be observed directly. Only the resulting emitted fragments were observable. What was needed was a high-density material in which tracks would be visible. The answer was photographic emulsions, since charged particles passing through an emulsion sensitize the grains of silver bromide with which they interact. At the University of Bristol in England Powell and Occhialini got together with scientists at the Ilford Company and produced emulsions much richer in silver bromide than ordinary emulsions. These emulsions not only provided them with much clearer tracks than the cloud chamber, but also allowed them to see exactly what happened at a collision with one of the silver atoms. Furthermore, the density of grains along the track gave a measure of the energy of the particle.

When Occhialini, Powell and C. M. G. Lattes exposed some of these new plates on mountain tops to cosmic rays they found totally new tracks. These tracks clearly showed a particle passing through the emulsion and coming to rest. Also from the point at which the particle came to rest another particle, clearly identifiable by its track as a muon, emerged. The pion had been discovered. Previously obtained results were much clearer now. The pions, were strongly interacting, and were produced much higher up in the atmosphere. Because their interaction was so strong, they interacted with the atoms in the air high up in the atmosphere and decayed into muons. The muons that only interacted weakly then penetrated much lower into the atmosphere. That was why it had taken so long to find the pions.

After Blackett and Occhialini obtained the first photograph of electron-positron showers from cosmic rays, Occhialini rushed over to Rutherford's house to show him the results. The following story circulated among his colleagues. Rutherford's maid opened the door and Occhialini in his excitement, kissed her. At any rate, what is fact is that Rutherford wrote Occhialini a check for fifty pounds, since at that time the young man was very short of money.

Occhialini and Powell also were the first ones to propose and verify a theory of pair production by a gamma ray passing close to a nucleus. This was the first time that the formula $E = mc^2$ was verified in the reverse direction and energy was used to produce mass, namely an electron and a positron.

The statement by I. I. Rabi, "Who ordered that" after the muon was shown not to be the Yukawa particle, was echoed several times. Later Murray Gell-Mann and E. P. Rosenbaum stated, "The muon was the unwelcome baby on the doorstep, signifying the end of days of innocence."

A startling surprise soon followed: nature distinguished between left and right at the most fundamental level. In pre-quantum days it was clear that any phenomenon that occurred, as viewed in a mirror, could also occur in the real world. There was no physical law that might distinguish between the world viewed in a mirror and the real world. It would be impossible to tell whether an observation was made in the real world or the mirror world. True, left and right are interchanged and we can tell whether a man's jacket is buttoned on the correct side, but then the jacket might have been made reversed to begin with. This apparent left-right asymmetry in clothing is not due to a law of nature. Most physicists believed that mirror symmetry, or parity, P was preserved in all fundamental interactions.

If we associate an arrow with the direction of motion of some object, then the arrow in the mirror is reversed; the image of an arrow pointing into a mirror, points out, except for an arrow associated with the motion of a screw. Consider a right hand screw. Its mirror image is a left-hand screw. The mirror has interchanged left and right. Now if we associate an arrow with the direction in which the screw moves when it is turned then, for a right hand screw facing the mirror, the arrow associated with this motion is into the mirror. The mirror image is a left-hand screw facing out of the mirror and the arrow associated with the motion due to the rotation is also into the mirror, that is, in the same direction as the original arrow. It is not back, out of the mirror. Physicists call such an arrow an "axial vector". There is nothing strange about such axial vectors, only that they do not reverse their direction when reflected in a mirror. Spin and angular momentum are both associated with rotation and are two examples of axial vectors.

The parity puzzle started in cosmic rays. Two particles, distinguished only by their decay products (the particles into which they decay) had been observed and were called Theta (θ) and Tau (τ). One of them decayed into two pions and the other into three pions. Since the pion carries an intrinsic negative parity, this meant that Theta and Tau had to have different parities. However, these particles were identical in every other respect. The simplest way to explain this was to assume that the two were really the same particle and parity was not conserved. So, sometimes the particle (that was Tau and Theta) decayed into a positive parity state and sometimes into a negative parity state. Two young Chinese-born physicists, Yang Chen Ning (1922 –) and Lee Tsung Dao (1926 –) suggested that there was no experimental evidence that parity was conserved for weak interactions and that this should be tested experimentally.

Now, all fermions, that is electrons, protons, neutrons, and neutrinos have spin 1/2. This is a rotation like a screw; it is an axial vector. For neutrinos, which have mass zero, the spin must (due to relativity) always be either parallel or antiparallel to the direction of motion. This is called the helicity of the particle.

In 1946, Wu Chien Shiung (1912 – 1997), joined the staff of Columbia University and became professor of physics in 1957. After reading the 1956 article by Lee and Yang she decided to test parity conservation. In an experiment at Columbia University, she aligned the nuclear spins in a sample of Cobalt 60. When Cobalt 60 beta decays to Nickel 60 it emits an electron and an antineutrino. In her experiment Madame Wu found that, for the Cobalt 60 nuclei with aligned spins, 40% more electrons were emitted in the direction of positive helicity (along the axis of rotation) than in the opposite direction. Mirror symmetry was not preserved. Parity (P) was not only violated, but violated as much as possible.

In 1963 she went on to experimentally confirm the theory of conservation of vector current in beta decay, a theory proposed by Richard Feynman and Murray Gell-Mann in 1958. She also observed that electromagnetic radiation from the annihilation of positrons and electrons is polarized, as predicted by Dirac's theory of the electron. In 1990 an asteroid was named in her honor.

Before Wu's experiment, Pauli did not believe that the attempt to detect parity violation would succeed and offered a very substantial bet. In fact, with reference to left-handed neutrinos he quipped, "Nonsense, God could not be weakly left-handed." After the experiment did indeed show that parity was not conserved he wrote, in a letter to V. F. Weisskopf, that he had been indeed lucky that no one had accepted his bet since he would have lost a substantial amount of money that he could not afford and as it was he had only lost some of his reputation which he could afford.

Sir Rudolf Peierls had asked Abdus Salam (1926 – 1996) on his Ph.D. oral examination (viva), "The photon has a zero mass because Maxwell's equations satisfy a gauge symmetry. So why is the neutrino mass zero?" This was an unfair question to ask of a graduate student since no-one knew the answer. However, a few years later, after hearing Yang lecture on his and Lee's theory of how the sacred principle of parity (left-right symmetry) may be violated in weak interactions, Salam remembered Peierls' question. He concluded that the answer to this question was that parity was violated and the neutrino satisfied instead a chiral symmetry. On presenting this answer to Peierls he was told, "I do not believe that left-right symmetry is violated in weak nuclear forces at all. I would not touch such an idea with a pair of tongs." Although discouraged, Salam did not give up and decided to ask the high priest of weak interactions, the father of the neutrino, Pauli. The answer that came back via Professor Villars was, "Give my regards to my friend Salam and tell him to think of something better." A few months later, after Wu had shown experimentally that parity was definitely violated in weak interactions, Pauli wrote a very apologetic note to Salam.

Salam came from a small town in what is now Pakistan. At age fourteen he received the highest marks ever recorded for the matriculation exam at the University of Punjab. When he returned home on his bicycle, the whole town had turned out to welcome him. He went on to Cambridge and obtained a BA (honours). A year later (1950) he received the Smith's Prize from Cambridge for the most outstanding pre-doctoral contribution to physics. His Ph.D. thesis on QED secured him an international reputation.

After a short stint of teaching at Lahore he returned to Cambridge as lecturer. In 1957 he was appointed Professor of Theoretical Physics at Imperial College, London. He went on to found the International Center for Theoretical Physics (ICTP) at Trieste. Many young physicists from developing countries have had an opportunity through the "Associateships" that Salam created to spend some time there in contact with active researchers.

Incidentally, Salam was once promoted to Brigadier General for one day in order to be able to fly back to London on an American Air Force Flight.

Lee and Yang and independently Lev D. Landau (1908 – 1968) had an idea that might mitigate the parity violation situation. They proposed that it was only the neutrino with its intrinsic helicity that was the cause of the problem. This would also explain why Cobalt 58, which decays into Iron 58 with the emission of a positron and a neutrino, emits the positron preferentially in the direction of negative helicity, opposite to that of Cobalt 60.

Landau was also very unhappy with parity violation until he realized that if particle and antiparticle are also interchanged, that is electron and positron, as well as neutrino and antineutrino then the symmetry is restored.

This interchange of particle and antiparticle is called "charge conjugation" (C). Thus, this "combined parity", as Landau called it, or CP symmetry was preserved and in a sense the classical idea of mirror symmetry again restored. This was indeed the case, but only for a few years.

In 1965 V. Fitch (1923 –) and J. W. Cronin (1931 –) of Princeton found that this combined parity was also violated in the decay of the strange particles K_0. In nature there is a definite distinction between left-right as well as particle-antiparticle. This may offer a clue as to why our universe, as far as we can tell, is made up entirely of matter instead of anti-matter.

At that time I was a graduate student in Princeton preparing for the General Examinations. My colleagues and I studied the Cronin-Fitch experiment and accompanying theoretical arguments in great detail since we were convinced that this would be a question on the exam. It wasn't.

After high energy machines, such as the cyclotrons and cosmotrons, came online it was possible to have beams of pions and in short order a host of new particles were observed. First there were the same particles K_0, K_+, K_- as had been observed in cosmic rays. Then there were also particles such as Lambda (Λ), Sigma (Σ), Xi (Ξ). All of these were initially called V-particles because of the characteristic V produced when they decayed. From a physicist's perspective, these particles were strange indeed. They were produced strongly but decayed only weakly, contrary to what quantum physics had to say. Abraham Pais (1918 – 2000) and Kazuhiko Nishijima (1926 –) cleared up the situation when they noted that the products of the decay were not the same as the products that produced these particles. In fact, these strange particles were always produced in pairs. This "associated production" lead M. Gell-Mann and independently T. Nakano and K. Nishijima to postulate, in 1953, that these particles carried a new quantum number that had to be preserved. In their usual imaginative manner, these physicists dubbed the new quantum number "strangeness". This meant that when a pair of these particles was produced one of them had the quantum number $+S$ and the other $-S$ (S for "strangeness") so that the total strangeness was 0, as it had been before the particles came into existence. This quantum number was conserved in strong and electromagnetic interactions and thus these particles could decay only via the weak interactions.

The following is from E. Andronikashvili's book *Reflections on Liquid Helium*. When Andronikashvili asked Landau, "Dau, what exactly is the physical meaning of strangeness?" the latter replied, "I don't know. And can you tell me exactly what isospin, which Heisenberg introduced in his day, is?"

Andronikashvili replied, "Let's assume that I don't know, but it's impossible that you don't know since you played such a part in getting it into scientific usage."

"I don't know. In my opinion it has no physical meaning."

Andronikashvili persisted, "How can it have no physical meaning when so many new real elementary particles have been discovered with its help?"

"Nevertheless, I insist that the concept of isotopic spin describes the degree of our lack of knowledge of the nature of elementary particles rather than a state of elementary particles themselves." To this Andronikashvili exclaimed, "Dau, you are an abstractionist."

Landau agreed, "I am an abstractionist," and ran along the corridor of the institute telling everyone, "Élevter has proclaimed me an abstractionist!"

In the early sixties, when publications in *The Physical Review* increased at an exponential rate C. N. Yang is supposed to have made the following statement. "*The Physical Review* is increasing so rapidly in size, that if this continues, the rate of advance of the leading edge of the magazine being piled up on the bookcases will, by the end of the century, exceed the speed of light. This will, however, not violate relativity since the information content will have gone to zero."

At the 1969 Conference on *The Physicists Conception of Nature*, Yang and Dirac urged physicists to abandon prejudices. They gave two guides as to what are, or are not, prejudices.

— Look at experiments.

— Look at the elegance of form and the simple beauty of logic.

In the 1960's, Murray Gell-Mann and Yuval Ne'eman (1925 –) were able to combine strangeness and isospin in one mathematical structure that used group theory. They called their approach the Eight-fold way, a clear reference to the Buddha. Actually the eight refers to the eight-dimensional representation of a mathematical group called SU(3). After that, one of the favourite tricks for particle physicists to play on reporters, when they wanted to take the physicists' picture with some complicated equation, was to stand beside the equation

$$3 \times 3 = 8 + 1.$$

This equation comes from group theory and says how the product representation of SU(3) on the left side decomposes into an eight-dimensional and a one-dimensional irreducible representation. So, it is a fundamental equation

for the eight-fold way. The three dimensional representations are the most fundamental and are carried by what the authors called "quarks", after the following verse in James Joyce's *Finnegan's Wake*.
Three quarks for Muster mark! Sure he hasn't got much of a lark. And sure any he has it's all beside the mark. But O, Wreneagle Almighty, wouldn't un be a sky of a lark.

The myth circulating among physics graduate students at the time was that the term "quark" occurred on page 383 of that book.

One of the predictions of this theory was the existence of a very strange particle the Ω_-, which had to have the unheard of strangeness of –3. In 1968 when this particle with the very large strangeness was indeed found, the eight-fold way was established. The Ω_- was predicted for the ten-dimensional representation of SU(3) and corresponds to the equation

$$3 \times 3 \times 3 = 10 + 8 + 8 + 1.$$

There was quite a strong, but friendly, rivalry between Feynman and Gell-Mann. Feynman came up with the "Parton Theory", which was a sort of generic form of quark theory without reference to Gell-Mann's quarks. This annoyed Gell-Mann so that he exclaimed, "Partons are stupid. Anyone who wants to know what the parton model predicts needs to consult Feynman's entrails."

In *A Festschrift for I.I. Rabi*[1], Gell-Mann had this to say with regard to the origin of the quark masses. "As to the discussion in the last section on the origin of the masses, I can only hope in the spirit of Rabi's remark, that whatever these speculations lack in artistry, they will make up in effrontery." In this same Festschrift, Abraham Pais stated, ". . . , it is often better to know the truth than to know the whole truth."

The experiments that provided support for the quark model and the eight-fold way were made possible by the development of the hydrogen bubble chamber by Donald Arthur Glaser (1926 –). Story has it that Glaser hit upon the idea while sipping beer and watching the bubbles rise.

Glaser confessed in an interview that when he finished high school and got to university he had no idea of the standards required since he had breezed through high school. On his first university physics exam he got a zero. So, when he went to see the professor and told him that he wanted to be a physicist, the reply was, "No way! To be a physicist you must get an A and

---

[1]New York Academy of Sciences, 1977.

with a zero on your test you can longer get an A." However the professor was willing to bend the rules and told him, "I'll make you a deal, even though it is against the rules. If you get 100% on the rest of your tests, I'll give you an A." Glaser got an A in that course.

In 1960, the same year in which he married Ruth Bonnie Thomson, Glaser received the Nobel Prize in physcs "for the invention of the bubble chamber."

Eight years later, Luis Walter Alvarez (1911 – 1988) also received the Nobel Prize in physics "for his decisive contributions to elementary particle physics, in particular the discovery of a large number of resonance states, made possible through his development of the technique of using hydrogen bubble chamber and data analysis"

Not content, Alvarez applied his knowledge to other areas. He later used cosmic rays as X-rays to search for hidden chambers in the pyramids. His greatest fame, however, came for his theory and experimental evidence for the extinction of the dinosaurs due to a catastrophic meteor collision with the earth. He was the first to notice a layer of iridium-enriched clay at the Cretacious-Tertiary (K-T) boundary [2]. This suggested that the impact of a meteor (meteors are rich in Iridium) may have been responsible for the extinction of the dinosaurs.

Alvarez believed in recognizing ability and made the following statement. "There is no democracy in physics. We can't say that some second-rate guy has as much right to an opinion as Fermi."

Later, in a series of experiments, that also lead to a Nobel Prize, parton theory was verified. One of the collaborators of this experiment was Richard E. Taylor (1929 –). At the conference, in Bamberg in 2001, to celebrate the 100th anniversary of the birth of Heisenberg the question of priorities of some discovery came up. I heard Taylor state, "It is not important who discovered this. It is more important that it was discovered." This should be compared to a statement by Eugene Wigner. "It is important to write the first paper on a subject, but it is more important to write the last."

At the same Bamberg conference, Taylor demonstrated that he is truly an experimentalist. While chairing a session he addressed everyone as "gentlemen". I pointed out to him that there were ladies present. After a quick perusal of the audience he commented that he was 99% correct.

At this same conference Harald Fritzsch was present. He told the following

---

[2]The cretacious period was from 136 to about 65 million years ago. The tertiary period was from about 63 to 1.6 million years ago.

story. When he was a student, Heisenberg asked him to prepare a talk on weak interactions and the intermediate vector boson, W. It took him several months to prepare and by then a paper by Murray Gell-Mann and Sheldon Glashow (1932 –) had appeared which predicted a W with a mass of some 70 GeV [3], at a time when everyone expected the mass to be about 2 GeV. So, at the end of his talk, Fritzsch mentioned this "funny" paper and everyone in the audience, except Heisenberg, laughed. After the talk Heisenberg asked Fritzsch to come to his office where he explained to him why this paper should be taken very seriously.

Several years later at a Coral Gables Conference, Eugene Wigner asked Fritzsch after his talk, "What about your theory, could it be wrong?"

Carlo Rubbia (1934 –) later was the principal investigator when the W as well as the Z particles were found. He believes that government funding for physics is inadequate and stated this as follows, "Physicists are like lemons; you squeeze them for all they are worth and then throw them away."

This is what Edward Mills Purcell (1912 –) had to say with regard to the utility of a pure science such as particle physics. "In our ignorance, it would be presumptuous to dismiss the possibility of useful application as it would be irresponsible to guarantee it".

At a conference when Wigner was already quite old, Gell-Mann was the last speaker of the session and went beyond the time limit, but the chair did not stop him. Wigner became more and more restless. Finally he stood up, pointed a finger at Gell-Mann and asked, "Who is this man?"

There is another story that may be purely gossip. Apparently, Gell-Mann had not been accepted into graduate studies at Princeton. So, in 1969 when he received the Nobel Prize, Wigner wanted to see why Gell-Mann had failed to be accepted. Unfortunately all records had been destroyed, but Wigner was able to find out who had written the letters of recommendation for Gell-Mann. One of these gentlemen was still alive and after Wigner wrote to him to ask what he had said he received an answer to the effect that, "I did not think much of him then and I see no reason to change my mind."

In 1974, Melvin Schwartz (1932 – 2006) [4] of Columbia University heard

---

[3]It is common for physicists to measure mass in units of energy. One GeV is a thousand million electron volts or about 10^{-24} gram.

[4]Melvin Schwartz shared the 1988 Nobel Prize in physics with Leon Lederman and Jack Steinberger "for the neutrino beam method and the demonstration of the doublet structure of the leptons through the discovery of the muon neutrino."

rumors that Samuel Chao Chung Ting (1936 –) at BNL [5]had discovered a remarkable new subatomic particle. When Schwartz confronted him — they were then working in the same laboratory — Ting flatly denied the rumor. Still suspicious, Schwartz offered to bet $10 that Ting had found the particle. Ting accepted.

Ting immediately wrote in his lab notebook, "I owe Mel Schwartz $10." Later Ting said he had been put in an "impossible position" since he wanted to check possible errors before making an announcement. He paid up just two months later, when he went public after learning that another group led by Burton Richter (1931 –) at SPEAR [6] had also seen the particle. Later Schwartz said, "What I keep kicking myself about is that I didn't make it $200."

Ting and Richter shared the 1976 Nobel Prize in physics "for their pioneering work in the discovery of a heavy elementary particle of a new kind."

The particle discovered by Richter and Ting is called the J/ψ particle. Richter had named it ψ and Ting had named it J. The J part of the name is almost identical to the Chinese character for "Ting". Gossip has it that Ting had named this particle after himself.

---

[5]Brookhaven National Laboratory
[6]Stanford Positron-Electron Asymmetric Rings

Chapter 8

Of Matter Liquid and Solid

" I should understand this better, but I don't." Victor F. Weisskopf, About Liquids.

Viktor Weisskopf (1908 – 2002) was thrilled when Pauli hired him to replace Hendrik Bugt Casimir (1909 – 2000) as his assistant. He could not understand why Pauli had chosen him over Hans Bethe, but Pauli cleared this up after Weisskopf arrived. "First I wanted to take Bethe, but he works on solid state theory, which I don't like, although I started it." Pauli was not just bragging since he was the first to apply quantum mechanics to crystals. In this work he explained the paramagnetism of alkali metals. Much later in life Bethe said with reference to this work by Pauli, "And that was important."

After Weisskopf arrived, Pauli then gave him a problem to work on. A week later he came to see what progress Weisskopf had made. After looking at the results he stated, "I should have taken Bethe after all."

Pauli's original comment was due to the following. For his Ph.D. thesis Sommerfeld had asked Bethe to use the Schrödinger equation to study the diffraction of electrons by crystals. In Bethe's words, he simply had to translate the thesis of Paul P. Ewald (1888 – 1985) on the diffraction of X-rays by crystals. This worked well and then he became much more ambitious and tried to improve the results. What came out was very complicated because of the much stronger interaction between electrons and ions. Thus, the final thesis was very long and to some extent messy. As a consequence, the first time Bethe met Pauli that worthy declared in typical fashion, "Mr. Bethe I was very disappointed in your thesis. From Sommerfeld's talk about you, I had expected better of you."

This was very much in keeping with Pauli's attitude to condensed matter physics. Once he even proclaimed, "One shouldn't work on semiconductors, that is a filthy mess; who knows whether they really exist."

Hans Bethe, especially in his youth, was not all work; he had his mischievous side. The January, 1931 issue of *Naturwissenschaften* contains a paper by G. Beck, H. Bethe, and W. Rieszler with the title, "Concerning the Quantum Theory of the Absolute Zero of Temperature." It was an elaborate hoax by these three postdoctoral fellows at the Cavendish Laboratory who pretended to derive the value of absolute zero temperature. In essence all they had done was to argue that temperature was related to degrees of freedom which were given by $(2/\alpha - 1)$ where α is Sommerfeld's fine structure constant and has the approximate value $1/137$. Hence absolute zero (that is $-273°C$) was supposed to be given by $T_0 = -(2/\alpha - 1)$.

The editor, was not amused and in the March 6, 1931 issue the pranksters published a "Correction" in which they stated: "The Note by G. Beck, H. Bethe, and W. Rieszler published in the January 9 issue of this journal, was not meant to be taken seriously. It was intended to characterize a certain class of papers in theoretical physics of recent years, which are purely speculative and based on spurious numerical agreements. In a letter received by the editors, from these gentlemen, they express regret that the formulation they gave this idea was suited to produce misunderstandings."

Regarding the fine structure constant, Feynman had this to say. "It's one of the greatest damn mysteries of physics: a magic number that comes to us with no understanding by man. You might say that the 'hand of God wrote that number, and we don't know how He pushed His pencil.' "

Rudolph Peierls (1907 – 1995), who had also worked with Pauli, advised Weisskopf that if he had to give a seminar then the best thing to do was to go Pauli the day before and present the talk to him in private. That way Pauli would be able to voice all his objections without destroying him in front of an audience. Then, unless he felt that Pauli's criticisms were valid, he could present the identical talk and Pauli would do no more than to mutter, "I already told him that."

Rudolf Peierls could also be quite strong in his opinions. When the paper on the Bohm-Aharonov effect appeared, J. S. Levinger (1921 -) happened to be in Oxford. Since he was then working on photo-disintegration, this paper interested him very much. So, he went to see Sir Rudolf Peierls to ask this eminent physicist for his opinion of this paper. The answer was, "This is utter nonsense. I don't see how the *Physical Review* could publish such tripe."

Several months later after experiments had verified the Bohm-Aharonov effect, Levinger again approached Sir Rudolf Peierls to ask his opinion on this result. This time the answer was, "It is obvious. There is no other way it could be."

On another occasion when a young physicist presented a talk in which he showed that a certain class of theories was impossible (the so-called "No-Interaction Theorems"), Sir Rudolf Peierls stood up after the talk and declared, "This is a very elegant proof of a trivial result."

Pauli, when he stated that he had started solid state theory was referring to his Exclusion Principle, which was later generalized to Fermi-Dirac statistics and is applicable to electrons and all particles with half odd integer spin. The other possible statistics, Bose-Einstein statistics, is applicable to photons and particles with integer spin.

In 1924 a young Bengali from Dacca University had sent a paper to Einstein. The *Philosophical Magazine* had previously rejected his paper and Satyendra Nath Bose (1894 – 1974) hoped that Einstein would see the value of his work and help him to get it published. The paper contained a totally original derivation of blackbody radiation. Einstein was most impressed with this paper. He not only translated it himself into German, but submitted it to the *Zeitschrift für Physik* with the following note. "In my opinion, Bose's derivation of the Planck formula constitutes an important advance. The method used here also yields the quantum theory of the ideal gas, as I shall discuss elsewhere in more detail." Incidentally the name "boson" for particles that satisfy Bose-Einstein statistics was coined by Dirac.

Although Pauli despised semiconductors, he lived to see their full impact with the development of the transistor. However, he does not seem to have changed his opinion. When John Bardeen (1908 – 1991), Walter H. Brattain (1902 – 1987) and William B. Shockley (1910 – 1989) invented the point contact transistor, they were on the way to make the vacuum tube obsolete. This opened the door for the second generation of computers. The first electronic computer, the ENIAC (electronic numerical integrator and computer), was invented by the US Ordnance Ballistics Laboratory during World War Two. Its purpose was to calculate trajectories of projectiles and at its best could calculate a sixty-second trajectory in about thirty seconds.

Today the ENIAC would be considered gargantuan in size and miniscule in computing power. It filled a large room, required 19, 000 vacuum tubes and used 200 kilowatts to run all these tubes and the air conditioning that kept everything from melting. The whole thing weighed more than thirty tons and kept breaking down as tubes burned out. Nevertheless it worked, with continual improvements, from 1944 until 1955 when its power was shut off for the last time. For its last six years it had consistently performed 100 trouble free hours a week of operating time.

Other computers soon followed. The one best known to physicists was MANIAC (Mathematical Analyzer Numerator Integrator and Computer). The acronym clearly shows that physicists were responsible. It was partly designed by Johnny von Neumann, who had also worked on the ENIAC. MANIAC was built at Los Alamos during the late 1940's and early 1950's and was used in the development of the hydrogen bomb.

However, computers based on vacuum tube technology were limited by their huge power requirements and size. The same thing was true of long distance telephone communication which also required signal amplification. Thus, it came about that Bell laboratories hired William Bradford Shockley to put together a team that might find a way to employ the semiconductor technology that had been so useful in developing radar technology.

Shockley was an athletic self-assured man. While attending Caltech he earned some extra money by modelling for *Telor Strength Equipment*. At the same time he exercised his penchant for practical jokes. One of his most famous was the case of the student "Helvar Scavi". He played this prank on professor Fritz Zwicky (1898 – 1974). Fritz Zwicky was a well-known astronomer physicist. In 1933 he had, on the basis of observation of the red shift in nebulae, introduced the concept of "dark cold matter".[1]

Professor Zwicky was known to ask students on their oral exams questions, other than those which he had indicated he might. Also, he did not pay attention to who attended his class. Shockley used these characteristics to advantage. He invented a fictitious student, Helvar Scavi, and enrolled him in Zwicky's class. Next he arranged that another student attend Zwicky's open book midterm exam. This student would slip his copy of the exam out an open window. Then, some of the most gifted students (legend has it that Robert Oppenheimer was one) who had already taken the course, answered all but one of the questions correctly. For the last question they wrote, "I'm too damn drunk to write any more." They then slipped the exam back to the fake student.

---

[1]See Chapter 9

After a year of such exams, Zwicky was astounded by the brilliance of this student, Scavi. However, in a sense, Zwicky unwittingly had the last laugh because he marked the students relative to each other. Scavi got an A and all the other students got a C.

On another occasion, several years later when Shockley was a recent hire at Bell Laboratories, one of his former teachers, Karl Darrow arrived to give a talk. As Darrow started to present his slides, a windup mechanical duck started to waddle across the stage. The laughing of the audience and the quacking of the duck caused Darrow to turn around and watch the duck continue its unperturbed waddle past him and off the stage. Although noone admitted to the prank, everyone knew it was Shockley.

At Bell Labs, Shockley assembled a team that consisted of Walter Houser Brattain, an experimentalist with the reputation that he could build or fix just about anything, John Bardeen, a theorist from the University of Minnesota, as well as various other physicists, chemists and engineers. The team worked well together and Shockley's secretary, Betty Sparks said that they had great parties and named their laboratory, "Hell's Bells Laboratory".

Shockley first came up with a possible design for a semiconductor amplifier in 1945. However, the device did not work and he asked Bardeen and Brattain to find out what was wrong. By 1947 they still could not make it work and Brattain decided to "baptize" the apparatus by dipping it in water. Lo and behold! It worked, not very well, but it worked.

Before the end of that year, Bardeen had one of those "Aha!" moments: electrons formed a barrier on the surface of the crystals they were using. Using this insight and without informing their leader, Bardeen and Brattain went ahead and built what was the first point contact transistor.

The name "transistor" was coined by John Pierce. Brattain had shown Pierce the device and explained how it worked. According to Pierce, "I thought right there at the time, if not, within hours, I thought vacuum tubes had *transconductance*, transistors would have *transresistance*. There were resistors and inductors and the name should fit in with the names of other devices, such as varistor and thermistor. And ... I suggested the name *transistor*."

Shockley was both pleased and upset when Bardeen and Brattain informed him of their success. He had wanted to be involved. Not to be outdone, he set to with a frenzy and, in a matte of four weeks of calculating — started in a Chicago hotel room — came up with a new more stable and rugged design. Two years later his junction transistor was ready for manufacture. However, the competitiveness broke up the team. Once the transistor was a fact, Shockley left Bell Labs to form his own company, *Shockley Semiconductor*. This was the beginning of "Silicon Valley" in California. Unfortunately there were personality clashes and some of Shockley's people left to found their own companies, in particular Bob Noyce and Gordon Moore founded Intel Corporation. Although not a single member of the original team of Bardeen, Brattain, and Shockley made much money from their invention, they shared the 1956 Nobel Prize in physics "for their researches on semiconductors and their discovery of the transistor effect".

Later Shockley developed some rather controversial theories about race and intelligence. This caused him to be viewed in a somewhat dim light by much of the physics community. Bardeen, on the other hand, had not yet achieved his greatest work. This was to come in 1957.

By the beginning of the twentieth century many gases had been liquified. There still remained one gas that proved most difficult: Helium. The same Heike Kammerlingh Onnes (1853 – 1926) that had thwarted Zeeman, was the first to do so in 1908. He was well prepared in physics since he had studied under Gustav Robert Kirchhoff (1824 – 1887) and Robert Wilhelm Bunsen (1811 – 1899) in Heidelberg.

There had been fierce competition among the various low temperature laboratories to achieve the liquification of this, the last gas to be liquified. Sir James Dewar (1842 – 1923), the first to liquify hydrogen, and the inventor of the vacuum flask, was in that race. He had an outstanding technician who was absolutely essential for running his laboratory. However, when Onnes succeeded before Dewar, the latter sent off a congratulatory telegram and then proceeded to his laboratory where he castigated his technician mercilessly. As a consequence the technician walked out of the lab and did not return until after Dewar's death. Without this technician very little useful work was accomplished in Dewar's lab.

Onnes laboratory was also the first example of large-scale physics. In 1911 he discovered superconductivity as well as the phase transition from Helium I to Helium II at the temperature of 2.172 K. All this required a whole range of new technical developments including means to measure temperature so close to absolute zero ($0 \text{ K} = -273.14°$ C).

As we already saw,[2] with regard to the Zeeman effect, Onnes was one of the great scientists of the Netherlands, but also quite the dictator. Although

---

[2]See Chapter 14 of *Quips, Quotes, and Quanta*

he died in Leiden the interment was in the nearby village of Voorschoten. All the technicians in his very large laboratory showed up in starched shirts with stiff celluloid collars, black coats and top hats. The funeral service started in Leiden with the funeral cortege following the horse-drawn hearse. Once outside the city limits, the horses speeded up. As a consequence the funeral cortege had to trot, at a good clip, for several kilometers. As one perspiring mourner gasped, "Just like the old man; even when he's dead he keeps you running."

Superconductivity remained a puzzle for a very long time. There were many wrong theories that claimed to explain it. This led Felix Bloch (1905 – 1983) in the late 1920's to enunciate what jokingly came to be known as a Second Bloch's Theorem. "Theories of superconductivity can be disproved." The actual, or First Bloch theorem is very useful and states that electronic waves in a periodic crystal have a specific type of wave function, now called a Bloch function. The second Bloch theorem was disproved in 1957 when John Bardeen, Leon Niels Cooper (1930 –), and John Robert Schrieffer (1931 –) published a theory (BCS theory) that explained the superconductivity of certain metals like copper, silver, lead, and tin. Their theory involves the pairing of electrons so that the pair acts like a boson. The really paradoxical thing about these electron pairs is that, although they are bound together, they may be physically far apart. This work earned the triumvirate a Nobel Prize in 1972. In the preface to his book *Theory of Superconductivity*, John M. Blatt makes the following interesting comment. "The idea that pairs of Fermions can combine to form effective Bosons was well known in chemistry, equally well known in molecular spectroscopy, and had, in 1931, proved of decisive importance in our understanding of the composition of atomic nuclei. Yet, the first published suggestion of a Bose-Einstein condensation of electron pairs as the cause of superconductivity was made in 1946, and then not by a theoretical physicist, but rather by an experimental chemist, R. Ogg!"

BCS theory led to Bardeen's second Nobel Prize in 1972. But before this came to pass, this eminent scientist got involved in a dispute with a twenty-two year old Cambridge graduate student, Brian Josephson (1940 –) and was shown to be wrong. Josephson got the Nobel Prize, one year after Bardeen's second Nobel prize. The twenty-two year old had published, in 1962 in *Physics Letters*, a paper that ostensibly infringed on the territory in which Bardeen was acknowledged master. In this paper Josephson presented an argument that showed that if an insulating barrier were sufficiently thin, a supercurrent would be able to tunnel through this barrier. Bardeen would have none of this and in an article of his own in the prestigious *Physical Review Letters* rejected this idea in a note added in proof. "In a recent note, Josephson uses a somewhat similar formulation to discuss the possibility of superfluid flow across the tunneling region, in which no quasi-particles are created. However, as pointed out by the author, pairing does not extend into the barrier, so that there can be no such superflow."

That Josephson was destined to do great work was evident early. When, in 1961 – 62, Philip Warren Anderson (1923 –) spent a sabbatical year in Cambridge he gave a series of lectures attended by Josephson. About this episode Anderson wrote, "Josephson had taken my course on solid-state and many-body theory. This was a disconcerting experience for a lecturer, I can assure you, because everything had to be right or he would come up and explain it to me after class." As a consequence of having become acquainted with Anderson, Josephson showed him his first calculations of the effect now called the "Josephson Effect". The calculations were clean and as Anderson said, "By this time I knew Josephson well enough that I would have accepted anything else he said on faith. However he himself seemed dubious." What puzzled Josephson and his supervisor, Alfred Brian Pippard (1920 -) as well as Anderson was that this tunneling supercurrent that Josephson predicted depended on the phase difference across the barrier.

Anderson shared the 1977 Nobel Prize with Sir Nevill Francis Mott and John Hasbrouck van Vleck "for their fundamental theoretical investigations of the electronic structure of magnetic and disordered systems".

That Josephson's work was far from trivial is revealed by an anecdote due to Felix Bloch. "Yang told me that he could not understand it, and asked whether I could. I had to confess that I could not either, but we made a deal that whoever of us first understood the effect would explain it to the other." This is the same C. N. Yang that had received the Nobel Prize for parity violation.

The dispute between Bardeen and Josephson continued in a direct confrontation in London. It must have been very intimidating for the young graduate student to stand up to a Nobelaureate and the acknowledged master of superconductivity. To both their credit, they presented their results in a quiet dispassionate manner. It was almost as if they were speaking a different language. Nevertheless, Josephson held his ground and asserted that Bardeen was wrong. When D. G. McDonald contacted him many years later for an article about this dispute (published in the July 2001 issue of *Physics Today*) Josephson replied in an e-mail, "Beware ye, all those bold of spirit who want to suggest new ideas."

In 1963 Anderson together with John Rowell published experimental proof of the Josephson effect. Ever the gentleman, John Bardeen conceded that Josephson's theory was correct. As already stated, in 1973, the year after Bardeen's second Nobel Prize, Josephson also got the prize.

Later Josephson became involved with ideas about mind, meditation and such matters and is now more or less at the fringe of the physics community.

My only personal contact with Josephson was when he visited the University of Alberta and we played a couple of games of Go. For a man already famous he was extremely modest and shy.

There was another puzzling phase problem that was discovered somewhat later by Sir Michael Berry (1941 –) and is named after him "Berry's phase". Quantum mechanics states that, in general, the phase of a wavefunction is not observable, only the difference of two phases is observable. Sir Berry found an exception to this rule, but still within the laws of quantum mechanics. With reference to Berry's phase, *Khagol*, the quarterly bulletin of the Inter-University Centre for Astronomy and Astrophysics of Pune, India, published the following sentence about the speaker at the 9th IUCAA Foundation Day. "It was a delightful experience for the IUCAA gathering to hear Sir Michael Berry about the phase that launched a thousand scripts."

John Bardeen was an avid golfer. Years after his second Nobel Prize, he was golfing with one of his friends. When he made a hole in one his friend asked which was better, a hole in one or a Nobel Prize. He replied, "I guess two Nobel Prizes."

Also, after John Bardeen received his second Nobel Prize in physics, C. N. Yang, the recipient of the Nobel Prize for parity violation, asked his colleague, Max Dresden, at Stony Brook, "Did you know that you could get the Prize more than once?" After receiving an affirmative answer he continued, "I guess I'll have to work harder."

Bardeen, Cooper and Schrieffer in obtaining their theory of superconductivity used a rather cumbersome technique. This was later greatly simplified by an approach used earlier in the Soviet Union by Academician, Nikolai Nikolaevich Bogoliubov (1909 – 1992) in the study of superfluid Helium. It is called a Bogoliubov transformation. Later, during his visit to the USA, Bogoliubov visited several campuses at a time when there was much agitation against the Soviet Union for its suppression of dissidents. As a consequence the Bogoliubovs were exposed to several verbal attacks. On returning to Rockefeller University, Bogoliubov confessed to Fredrick Seitz, "If I went back home and said to my government all of the things that the people at the other universities asked me to say, I would promptly be thrown in jail."

The superconductors so far discussed are all that was known pre 1986. Most physicists believed that it would be impossible to find a material that would become superconducting at temperatures above some 20 K (–253° C). There were even some (incorrect) theorems that proved that higher temperature superconductors were impossible. In 1986 J. George Bednorz (1950 –) and K. Alexander Müller (1927 –) surprised the physics world by showing that a ceramic, an insulator, became superconducting at the unheard of high temperature of 33 K (−240° C). Not only did they receive the Nobel Prize, but they set off a flurry of activity that has resulted in ever higher critical temperatures T_C , at which these new materials become superconducting. One of the most recent records is 135 K, or –138°C.

Most interesting is the fact that so far (2006), after almost two decades of intensive research, no theory of high T_C superconductors exists. The theoretical, as well as the experimental situation is still unclear. Thus, it is not surprising that when one of the researchers in the field, was giving a talk on the subject he was asked, "When will the premier experiment regarding the gap in high T_C superconductors be done?" He answered, "The premier experiment has been done but, we all don't agree on which it is."

The view about the lack of solution for high T_C as well as other condensed matter problems is according to David Thouless (1934 –) as follows. "Most problems in condensed-matter physics are unsolved, sometimes because they are too complicated, and sometimes because no one has seen the simplifying pattern. Obvious guesses about a big breakthrough in understanding glass formation or protein folding will probably be wrong. On the experimental side, the search for a room-temperature superconductor is an exciting challenge."

Besides superconductivity, Kammerlingh Onnes had made another great discovery with liquid Helium: superfluidity. Much of the early work on superfluid Helium was performed in the Soviet Union by Piotr Leonidovich Kapitsa (1894 – 1984), Lev Davidovich Landau (1908 – 1968), Élevter L. Andronikashvili and others. Thus, it is not surprising that Richard Feynman began his talk at the *Sixth International conference on Low Temperature Physics* in 1958 in Leiden with the following words. "I go from conference to conference in order to meet Professor Landau, and I am grieved that I can't find him anywhere. First of all because I am working on liquid Helium for which he has done so much, and I would like to talk to him about it. But also because every time I have to give the lecture that it was intended that he should give."

That, not only liquid Helium I, but also liquid Helium II is a superfluid, was discovered by Kapitsa. He had worked in Cambridge from 1921 to 1934, first with Rutherford and later in the Mond laboratory that had been built for him. Sir Baldwin, former and again later Prime Minister and chancellor of the University, opened the new Mond laboratory for Kapitsa in 1933. Kapitsa showed Sir Baldwin around and explained how things work and the special design that prevented the roof from blowing off in an explosion. At one point Sir Baldwin asked, "Is that so?" Kapitsa replied, "Oh yes, you can believe me. I'm not a politician."

Kapitsa was also responsible for a large painting of a crocodile on the wall of the Mond laboratory. He admitted that the crocodile represented Rutherford. "In Russia the crocodile represents the father of the family." He also said that the crocodile cannot turn its head. Like science, it must go forward with all-devouring jaws.

After his return to the Soviet Union, Kapitsa was detained there. To induce him to continue to do his extraordinary research, an institute was created for him in Moscow. Even so, Kapitsa depressed and unable to work, complained to Premier Molotov, "Don't you know that a bird in a cage doesn't sing?" To which Molotov replied, "This bird will sing." And sing the bird did. Eventually Kapitsa succeeded, through the intervention of Rutherford, to convince the British government to sell the equipment from the Mond Laboratory to the Soviet Union. Later when a British colleague complained regarding the brain drain of British scientists to the USA, Kapitsa commented that the USSR was in a more difficult position, having no one to drain.

Kapitsa was almost fanatical about neatness in his laboratory and that of his assistants. He was often heard to say, "Remember, where there's dirt there can't be good scientific results that a scientist could guarantee." He knew, that extremely low temperatures would lead to new discoveries. Accordingly, many of his lectures began with, "If you want to study some properties of a material in depth or discover new phenomena in it, then you must subject it to extreme conditions — conditions under which phenomena that might hinder you are either excluded or suppressed."

When Kapitsa was awarded the title Hero of Socialist Labour and was invested with the Order of Lenin and the Gold Star he planned a large reception in the institute of which he was the head. On the invitations he wrote, among other things, "Dress, formal; orders, medals; temperature in the room 19° Celsius" because he wanted, "to be in the best houses and ladies should not dream of coming in décolleté" since it was during the war and, to save energy, it was not allowed to heat rooms above 19° C.

The other great Soviet physicist that worked with devotion with Kapitsa was Lev Davidovitch Landau. His friends used "Dau", the nickname he preferred. He himself joked that his name came from the French spelling of Landau, namely, L'âne Dau, meaning "the ass Dau". As early as 1929, when he was only 21, Landau had solved the problem of diamagnetism in metals, something that had already produced a host of wrong papers. Pauli recognized the correctness of this work, although many theorists, due to all the earlier mistaken results, were skeptical. Dau's brilliance was thus clearly evident early in his career and he was sent to Western Europe for further study. On his first visit outside the Soviet Union, the Swiss government would not allow him an extended stay of a few months and issued a visa for only a very short period. To this Landau commented that Lenin had lived in Switzerland for many years without starting a revolution and now they were afraid that he would start a revolution in a matter of months.

During this first visit outside the Soviet Union Dau had the opportunity to work with several people, in particular with Rudolf Peierls (later sir Rudolf Peierls). The latter stated years later , "In those days we all lived off crumbs from Landau's table." From this period Peierls also related, "One of my favourite recollections is of when a physicist was mentioned, in the course of discussion, of whom Landau had not previously heard. He asked, 'Who's that? Where is he from? How old is he?' Someone said, 'Oh, only twenty-eight.' To which Landau exclaimed, 'What, so young and already so unknown?' "

Also, on this first trip outside the USSR Dau met a colleague from the Soviet Union, Yurii Borisovich Rumer (1901 – 1985), in Berlin and lamented to him, "All the pretty girls are already married, and all the nice problems are already solved." Years later Landau and Rumer co-authored a small popular book *What is Relativity*. Landau's comment on this book was, "Two rogues persuade a third that for ten Kopecks he can understand relativity." He seems not to have had much use for reading books on physics, "Nothing new can be learned from weighty tomes. They are a graveyard for past ideas that have served their time."

According to Hendrik Casimir, the following occurred in Zürich and shows not only Dau's disdain for books but also his faith in revolution. They were in a library where a beautifully bound collection of the French Académie des Sciences was prominently displayed. Dau taunted, "Let's have a look to see what these old idiots had to write. This should be fun." After he opened one volume to an important paper by Legendre, one by Lagrange, another one by Laplace and so on he replaced the books quietly and after a long period of silence his face lit up. "This shows how much the French revolution did for scientific progress." Clearly he forgot the fact that the revolution also beheaded Antoine Laurent Lavoisier. This execution had led Lagrange, to remark, "It took them only an instant to cut off that head, and a hundred years may not produce another like it." Landau would, however, have been amused by another statement by Lagrange. "I have always observed that the pretensions of all people are in exact inverse ratio to their merits; this is one of the axioms of morals."

In spite of his disdain for books, Dau and his lifelong collaborator, E. M. Lifschitz (1915 – 1985) wrote a nine volume series of texts on a course of theoretical physics. These books became standard fare for all theoretical physics graduate students and were used to educate more than one generation of physicists. However, Landau never physically wrote his papers himself. He usually dictated the paper to his assistant Lifshitz of whom he said, "Evgeny is a great writer; he cannot write what he does not understand." This meant that the writing had to be clear to the reader. The fact that the physical writing was done by Lifshitz and the dictation came from Landau led to the waggish, but unfair, comment that these volumes, "contained not one word from Landau and not one idea from Lifshitz."

While Landau was in Berlin, Einstein gave a talk to the German Physical Society, on a new idea he was working on. During the question period a young man at the back of the hall got up and in broken German said, "What Professor Einstein said is not so stupid, but the result of the second equation does not follow from the first. One has to make another assumption and furthermore the result is not invariant." Everyone in the hall was shocked by this incredible behaviour except Einstein, who stood facing the blackboard lost in thought. After some time he turned around and declared, "The young man is perfectly right. You can forget everything I said." The young man was Landau.

In 1930 Landau was also in attendance at the famous seminar in Berlin. The front row was occupied by the luminaries: Einstein, Planck, Nernst, etc., a truly intimidating group. At one point a distinguished-looking gentleman stood up and made some remarks. Landau jumped up and in broken German said, "I don't know who the gentleman is, but what he says is pure rubbish." The gentleman in question then rose again, and with a bow towards Landau introduced himself, "von Laue."

Landau also made the requisite visit to Niels Bohr's institute and there met Pauli, Teller, and other physicists. This may be what prompted much later the following. I. M. Khalatnikov mentioned to Landau how much he admired his critical intellect. To this the latter replied, "You have not met Pauli! That was a great intellect!"

The informality of the seminars at the Bohr Institute can be gauged from the following episode. During one such seminar, Landau and Bohr got into a discussion and Landau, who had worked late the previous night, was tired. So he stretched out on a bench and from this position continued the discussion with Bohr who was bent over him. Apparently neither one saw anything wrong with these positions in spite of the presence of a large audience.

On another occasion, Landau after a heated debate asked Pauli if he thought that all he had said was nonsense. This kind gentleman replied, "Oh no, far from it. What you said was so confused, one could not tell if it was nonsense."

As mentioned above, Landau and Teller were both at Copenhagen at the same time. At that time Teller got married and Landau asked him and his wife how long they planned to stay married. They told him for a very long time and had no plans whatsoever to ever terminate the marriage. Landau responded with strong disapproval and stated that only a capitalistic society could induce its citizens to take a perfectly good thing and spoil it by exaggerating it to such an extreme.

Unlike most great physicists, Landau never liked music and had no appreciation for it at all. At one of the Copenhagen conferences a number of physicists were invited to the Bohr home. Among these were Landau, Dirac and Casimir as well as Frisch, Kopfermann and Weisskopf. The latter three gave a performance of music and the others together with their wives sat and enjoyed the performance. All except Landau. This gentleman made no bones about being bored by this whole business: grimacing, waving his hands up in the air and generally bothering everyone. During a lull in the performance Dirac approached Landau and suggested, "Landau, if you don't like the music, why don't you leave the room?"

Landau responded, as was his nature, in a somewhat strong manner, "Well, I wanted to leave the room and Mrs. Casimir is also not interested in music and I asked her to leave with me, but she didn't. If she had left so would I. Why didn't she go out with me?" Dirac's response was gentle and quiet, "Obviously, she preferred to listen to the music to going out of the room with you." It was a memorable occasion for Landau had no answer.

On the other hand, Dau loved poetry and frequently recited long poems from memory. His poems had to have a definite rhyme and rhythm. Poets were aware of his fondness of their work and asked him to clarify the role of poetry in his life. He answered, "What is the use of poetry? A question as hard to answer as, 'What is the use of love?' For someone who loves poetry, it brightens and adorns life. For myself, without my favourite poems that I can repeat to myself whenever I wish, something would be missing. My favourite poet is Lermontov. How to write good poems is, of course, not something that can be explained theoretically, otherwise everyone could write marvellous poetry. Only a true poet can excite the reader." These statements should be compared to Dirac's question to Oppenheimer. "How can you do both physics and poetry? In physics we try to explain in simple terms something that nobody knew before. In poetry it is the exact opposite."

Landau's first position was at the Ukrainian Polytechnical Institute. When he arrived, and found on the door of his office a sign

PROF. L. D. LANDAU

he immediately added

BEWARE, HE BITES!

In 1938 Landau was arrested due a false accusation. The very day of the arrest Kapitsa wrote a letter to Stalin stating that neither Soviet nor world science could afford the loss of Landau. After a year of no reply he wrote to Molotov and again repeated in very clear terms how necessary it was to have Landau to help with the theoretical work on superfluidity, which Kapitsa had discovered in 1938. He even dared, with incredible courage, to criticize the Commissariat for Internal Affairs. Landau was released under Kapitsa's personal guarantee. In 1941 Landau vindicated Kapitsa's faith in his abilities and published a theory of superfluidity. This involved the idea that there were two components, a normal fluid and a superfluid. For this he was awarded the Nobel Prize in 1962. Out of loyalty and gratitude, Landau never left Kapitsa's institute. That is why Landau did not object, even when Kapitsa used to say, "Ask a theoretician and then do the opposite." His response was simply, "He saved my life, so I can't feel offended."

When E. L. Andronikashvili was trying to verify Landau's theory of superfluidity he ran into paradoxical results while measuring the viscosity of Helium. Dau was down in Andronikashvili's lab several times a day asking for the experimental results which seemed to yield a different value every time. The measurement was performed by rotating a disc suspended from a quartz wire in liquid helium and measuring the drag by means of the damping of the rotational vibrations. One day Dau shouted, "Fools!" and handed a formula to Andronikashvili to apply to his experimental data. When the latter asked how he had obtained it, Dau replied that he had used a well-known result from diffraction theory. To the question, "What does diffraction theory have to do with it?" he simply grinned. Then, the next day he announced to everyone in the institute that he had produced a simple correction formula for the edge of a disc oscillating in a viscous fluid by using Fraunhofer diffraction theory. He added, "You know, it's really quite simple." Everyone agreed, although no-one was able to reproduce his result and Dau refused to reproduce his derivation which had been done on a scrap of paper and either discarded or lost. After that all the mysterious results were cleared up and Landau's original theory was beautifully verified. Physicists from all over the world continued to use Landau's correction formula for the next twelve years, without knowing how it was obtained, until a Belgian physicist rederived it by a totally different technique. This was a typical example of Dau's intuitive approach to physics problems. It also explains the following quote, "How can you solve a problem without knowing the answer in advance?"

Later Andronikashvili, in further experiments, found that Landau's theory could not explain some completely new effects. These difficulties were resolved by a new theory due to Richard Feynman who showed that in liquid Helium II the superfluid component develops vortices. These vortices are quantized in the same way as Bohr orbits. However, in this case the orbits are not of atomic dimensions but are of macroscopic size of the order of centimeters. Again all this was verified experimentally.

Landau built up a magnificent school of physics. To teach his students about the most recent research they were required to participate in a seminar in which students, chosen at random, would report on papers selected by Landau. One of these students, V. Levich failed to appear to present his report. On the following occasion Landau was already in obviously high dudgeon. Levich arrived, approached Landau and handed him a note before the reprimand could commence. Landau read the note and burst into laughter. The note was a formal certificate with appropriate signature and seal that Levich was dead. Peace between teacher and student was restored.

At the Landau seminars Arkadii B. Migdal (1911 – 1991) always showed up at the last minute and it became a standard ritual for Landau to say to everyone who had already assembled at five or more minutes before the start, "There is still one minute left. Let us wait, maybe Migdal will come." Sure enough, at one minute to the hour the door would open and Migdal would enter.

At the celebration of his fiftieth birthday Landau said, "Some people think that a teacher steals from his pupils, others that pupils steal from their teacher. I think that both are right, and it is great fun to participate in this mutual theft."

Igor E. Tamm instructed his students on how to approach Dau with the following words, "Pay no attention to Landau's general remarks such as 'Rubbish' and 'Impossible!' but, when he starts to say something specific about your work, listen carefully and keep your wits about you."

Landau could also be wrong. He rejected irreversible thermodynamics with the words, "the thermodynamics of irreversible processes is irreversible rubbish". Another example is that he did not believe that plasmas constituted a fourth state of matter and stated, "There are three states of matter: solid, liquid and gas, and there is no fourth state and there cannot be".

What follows are some disconnected quotes and stories.

This is an illustration of Dau's sense of humour. On one occasion his landlady provided some home made cottage cheese. Dau looked at it and said, "Lucky I don't like this, because if I did I should eat it, and it tastes horrible".

Dau frequently stated that "the extermination of bores is the duty of every decent human being." Whenever he encountered a "bore" he would tease the individual mercilessly.

When Dau found out about the death of Fermi he was silent and obviously disturbed. Finally he said, "The fact is that anyone who loves physics cannot but deplore the death of Fermi."

Landau disapproved of publicity seeking in science. "People who have heard of some unusual phenomenon (in science or in life) begin to put forward implausible hypotheses to account for it. Look at the simplest explanation first, namely that it is all nonsense." This attitude would have saved many hundreds of incorrect "explanations" of room temperature fusion.

That Landau had achieved much in physics is well summarized by Isaac Y. Pomeranchuk (1913 – 1966), "You cannot image what a vast amount of sewage Dau cleaned out of theoretical physics."

In 1962 Landau had a tragic car accident. He was twice clinically dead but was kept alive. A Canadian neurosurgeon was specially flown to Moscow and managed to save his life. Unfortunately Landau was never his old again and died six years later. While in the hospital, recovering from the car accident that destroyed a portion of his brain, he was visited by his colleague Pomeranchuk. After Pomeranchuk asked him if he was the same as before Landau replied, "I am not as smart as Landau, but I am smarter than Pomeranchuk."

In his last published paper Landau wrote with regard to a more rigorous proof of his results that, "The brevity of life does not allow us the luxury of spending time on problems which will lead to no new results."

More than once, nature had a surprise for those who knew what to expect from a particular measurement. It was like Fermi said, "If you find agreement with theory you have made a measurement, but if you are very lucky you will find disagreement. Then, you have made an experiment!"

When Klaus von Klitzing (1943 –) wanted to measure the Hall resistance for materials that had electrons confined to a plane (so-called two-dimensional electrons) and for very strong magnetic fields he was told that this was a waste of time since there was nothing new to be learned from this. He disagreed since he knew that the two-dimensional electron density was independent of the material in which the electrons were held and therefore there had to be something universal. So he did his experiments and discovered the *Quantum Hall Effect* for which he received the 1985 Nobel Prize.

This is reminiscent of Stern's measurement of the magnetic moment of the proton, where all the people knew what the outcome would be and had argued that the measurement was a waste of time. They also were all wrong.

It also reminds one of a statement by Walther Gerlach (1889 – 1979), "No experiment is so stupid that it should not be tried."

Funding agencies should note these examples since they are most unlikely to fund an experiment that is not in the mainstream of research.

Condensed matter physics has probably had the biggest technological impact of any scientific endeavour in the last fifty years. Even so, physicists are still fairly low on the list of important people. The two stories that follow may show how, in spite of their contributions to society, physicists rank in society.

In the 1950's a group of Canadian physicists was travelling to the USA to attend a conference. At the border the immigration officer asked the usual question, "What is the purpose of your visit?" The response, "We are going to attend a physics conference" caused the officer to look them up and down and comment, "You don't look like athletes."

On a flight to Oklahoma City, three physicists, Lee du Bridge, Julius A. Stratton, and William V. Houston, all of whom later became presidents of their respective universities, were on the same plane with the crew of the Today Show. Their plane had a stopover of half an hour in St. Louis for refueling. The three physicists together with some other passengers decided to pass the half-hour by taking a stroll through the airport. When they returned to the gate their plane had departed ahead of schedule. Dave Garroway and the crew of the Today Show had used their celebrity status to convince the pilot to take off as soon as refueling had been completed. The scientists were able to catch a later flight.

Chapter 9

Cosmology

"..., remember that in the whole universe there is no lower limit, and that the atoms have no place where they may rest together, since space is without end or bound and extends without measure on all sides and in every direction." Lucretius, De rerum natura, ca 55 BC.

Every religion has some form of creation story, or cosmology. It seems that we humans want to know not only about our origins, but also about the origins of the universe as a whole. We want to be put in context. As Allan Sandage (1926 –) said about cosmology, "Why is there something instead of nothing? In science you can ask questions 'what?', 'how?', and 'when?'. The question 'why?' belongs to the realm of philosophy. I'm a scientist tonight so that question will surely not be answered. Yes, the evidence for the Creation Event is there. Yes, the Hubble constant is 50. No, not everybody agrees ...Progress in this field only comes at the funerals for the astronomers."

Steven Weinberg justifies cosmology as follows. "The effort to understand the universe is one of the very few things that lifts human life a little above the level of farce, and gives it some of the grace of tragedy." He also stated with regard to religion, "With or without religion, you would have good people doing good things and evil people doing evil things. But for good people to do evil things, that takes religion."

It seems that cosmologies are culturally conditioned. Thus, the early Egyptians viewed the cosmos as half of a cylinder that stretched above the Nile. Hindus had a giant turtle floating through the void. So, it went. Every culture had its own model of the universe. Today we have our own

scientific model of the universe. We should, however, remember the words of
Manfred Eigen (1927 –) as stated in *The Physicists' Conception of Nature*.
"A theory has only the alternative of being right or wrong. A model has a
third possibility: it may be right, but irrelevant."

This is what Steven Weinberg (1933 –) had to say about cosmology, "The
more the universe seems comprehensible, the more it also seems pointless."
Similarly Peter G. Bergmann (1915 – 2002), declared at the 4th Soviet Gravitational Conference, Minsk, USSR, July, 1976. "A physical theory remains
an empty shell until we have found a reasonable physical interpretation."
Eddington, on the other hand appears poetic. "We have found a strange
footprint on the shores of the unknown. We have devised profound theories, one after another, to account for its origins. At last, we have succeeded
in reconstructing the creature that made the footprint. And lo! It is our
own." Also, "Not only is the universe stranger than we imagine, it is stranger
than we can imagine." Sir James Jeans also seemed to echo Dirac, "From
the intrinsic evidence of his creation, the Great Architect of the Universe
now begins to appear as a pure mathematician." And on another occasion,
"...nature seems very conversant with the rules of pure mathematics, as
our mathematicians have formulated them in their studies, out of their own
inner consciousness and without drawing to any appreciable extent on their
experience of the outer world."

Modern cosmology is based on a series of astronomical discoveries starting
roughly in 1912 when Henrietta Leavitt (1868 – 1921) opened the doors to
measuring interstellar and later intergalactic distances, after she discovered
certain stars called Cepheid Variables. She examined these individual stars
in a distant galaxy and found that the brightness of these stars varied in
a regular fashion such that there was a definite relationship between their
luminosity and the period of their brightness variation. Since these stars
were all in the same distant galaxy and therfore almost the same distance
from earth, their variation in brightness was due solely to their variation in
luminosity and not at all to varying distances from us. Thus, by measuring the intervals between maximum brightness, one could know the actual
brightness of these stars and by further measuring the observed brightness
one could easily calculate the distance to these stars.

Leavitt had volunteered at the Harvard College Observatory in 1895.
Seven years later she was appointed to the permanent staff (at a salary of 30
cents an hour) by director Charles Pickering.

The next significant event occurred in the period between 1912 and 1915.

V. M. Slipher (1875 – 1969) determined the velocity (on a line from us) of 13 galaxies and found that all but two of them were moving away from us at about 300 km/s.

Einstein's General Relativity next opened up the possibility of studying the cosmos as a whole in a new way. He himself set the stage in 1917 by looking for a solution that would yield a universe in accord with the beliefs of his time that the universe was "unchanging and enduring from everlasting to everlasting". It would also answer the question that Newton had worried about. Why did the universe not collapse under its own gravitational attraction? Since Einstein's field equations seemed not to provide such an unchanging model solution, he modified his equations in the only mathematically permitted fashion and introduced a "fudge factor", the cosmological constant. This allowed him to find such solutions.

In 1917, Willem de Sitter (1872 – 1931) also found solutions of the Einstein equations (using the cosmological constant). These satisfied the requirements of an unchanging and isotropic universe.

In 1922, Aleksandr Aleksandrovich Friedmann (1888 – 1925) derived a model universe that evolved and remained homogeneous and isotropic. this universe could not be static.

The date of Friedmann's birth is often given as June 29. However, this is due to two errors which came about in converting from the old Russian date (Julian calendar) to the new date (Gregorian calendar). To do this requires that one adds 12 days. Rather strangely, especially for a relativist, Friedmann converted his own date of birth incorrectly to June 17. Actually it should have been June 16 since his old style birthdate was June 4. Now for the final blunder. Since it was not known that his birthdate had already been converted, the date was converted once more to June 29.

One of the greatest astronomers of the first half of the twentieth century was Edwin Hubble (1889 – 1953). He was born in Marshfield Missouri and later studied at the University of Chicago and Oxford. For all of his professional life (from 1919 until his death) he was on the staff of the Mount Wilson and Palomar observatories. One of his peculiarities was that he did not like his origins. So he reinvented himself and, to further distance himself from these origins, did not even attend his mother's funeral.

Finally, in 1929 after almost a decade of measurements, Hubble established the law named after him and showed that the universe was not static, but rather was expanding. He verified this law by observation out to a distance of six-million light years. Hubble recommended that the cosmological constant be dropped from the Einstein field equations. Earlier, in 1923, Hermann Weyl had also pointed out that the de Sitter model would yield an expanding universe according to what later became known as Hubble's Law.

When Einstein learned of the expansion of the universe, he happily got rid of this unwanted cosmological constant. He called the introduction of the cosmological constant, "the biggest blunder of my life". Willem de Sitter had also realized this and, regarding the cosmological constant, written in 1919 that "the term ...detracts from the symmetry and elegance of Einstein's original theory, one of whose chief attractions was that it explained so much without introducing any new hypothesis or empirical constant."

Today, the cosmological constant is back in a big way as "dark energy".

In 1927 Georges Lemaître (1894 – 1966) independently derived a model universe that evolved and remained homogeneous and isotropic. This derivation also included an expansion rate that showed that the rate at which galaxies receded was proportional to the distance (Hubble's Law) and Lemaître became known as the "father of the Big Bang Cosmology" Einstein did not accept this work since he wanted a static universe. He wrote to Lemaître, "Your mathematics is superb, but your physics is abominable."

Although a cosmologist, Lemaître was also a staunch Catholic. This is what he had to say about the biblical creation story (my translation). "There is no reason to reject the bible because we now think that it has perhaps taken ten billion years to arrive at that which we believe to be the universe. Genesis tries to simply teach us that one day a week should be dedicated to rest, to the prayer necessary for our health."

In 1932 Einstein and de Sitter published a joint paper with a particularly simple solution of the field equations of general relativity for an expanding universe. They argued in this paper that there might be large amounts of matter which does not emit light and has not been detected. This is what is now called "dark matter". Observations, due to its gravitational effects, have now shown that dark matter very likely exists; in fact that it constitutes 22% of all matter, dark energy constitutes 74% of matter and ordinary particles and energy constitute a mere 4% of all matter.

Regarding the size of the universe, Einstein had this to say, "Only two things are infinite, the universe and human stupidity, and I'm not sure about the former."

One of the (so far unobserved) predictions of the field equations of Einstein's General Relativity is the existence of gravitational waves. These were predicted early by approximate calculations using Einstein's equations. However, at one point Einstein almost published a paper negating the existence of these waves.

After moving to the United States, Einstein published a few papers in the *Physical Review* until 1936. From then on he never published another paper in that journal after an anonymous referee gave a negative criticism of a paper that Einstein and Nathan Rosen (1909 – 1995) had submitted.

While he resided in Europe, Einstein had published mainly in German journals. In these journals the editor decided on whether a paper should be published and only rarely had someone else referee the paper. In the USA, the *Physical Review* used an anonymous refereeing system. Einstein was furious that his paper had been refereed and that the referee had dared to claim that the paper was wrong. The interesting fact is that the referee was right and Einstein was wrong. Einstein's equations predicted gravitational waves, but in the submitted paper *Do Gravitational Waves Exist?* he and Rosen arrived at the incorrect conclusion that gravitational waves do not exist.

Einstein had written (regarding this paper) to Max Born, "Together with a young collaborator, I arrived at the interesting result that gravitational waves do not exist, though they had been assumed a certainty to the first approximation. This shows that the non-linear general relativistic field equations can tell us more or, rather, limit us more than we have believed up to now."

As the referee pointed out, Einstein's mistake was in assuming that a singularity (a point where the solution becomes meaningless) in his solution with Rosen was due to the physics, when in fact it was due to the coordinates that they happened to choose for writing the solution.

Einstein withdrew the paper and never published in the *Physical Review* again. He submitted the paper to the *Journal of the Franklin Institute* where it was published with a major revision that concluded that gravitational waves do exist. The *Journal of the Franklin Institute* had originally accepted the paper without alterations, but soon after Einstein explained to the editor that "fundamental" revisions were required because "consequences" of the equations obtained in the paper had been incorrectly inferred. There was no acknowledgement of the *Physical Review's* referee's contributions.

The theoretical developments involving astronomy and General Relativity ended one phase of the evolution of modern cosmology. The next phase involved some nuclear physics.

In the summer of 1938 Bethe finished his paper on the carbon cycle for energy production in the sun and thus in stars. He sent it to the *Physical Review*. However, this paper did not appear until 1939 and then in the *New York Academy of Sciences*. This came about as follows.

In 1938, I. I. Rabi sent Robert E. Marshak (1916 – 1992) as a graduate student to work with Bethe. Marshak pointed out to Bethe that the *New York Academy of Sciences* was offering a $500.00 prize for the best unpublished paper on energy production in stars. Bethe offered Marshak a 10% finder's fee if he won. He then withdrew his paper from consideration by the *Physical Review*, submitted it to the *New York Academy of Sciences* and won. Of this money he paid $250.00 to the German government to release his mother's furniture so she could take it with her when she emigrated.

Bethe was incredibly hard working with an unbelievable ability to concentrate. As professor at Cornell, he showed up every morning punctually at 9:00 AM, picked up his pencil and continued on exactly where he had left off the previous day. At the stroke of noon he put the pencil down and headed to the cafeteria where the students and postdoctoral fellows had reserved a table with a seat for him so that they could ask him questions. At 1:00 PM he was back at his desk and continued, without pause, where he had left off until precisely at 5:00 PM. The only exception was on Fridays when he attended the colloquium at 6:00 PM and sat at the very back with his eyes closed until he had some question.

When Victor Weisskopf intended to start a calculation of pair creation of Bose particles by photons, he went to Bethe for advice. The latter had recently published his famous paper with Heitler on electron-positron pair creation. As usual Bethe gave the advice freely. When asked how long such a calculation might take, Bethe replied, "It would take three days for me and three weeks for you." The estimate proved correct and even so the published result was wrong by a factor of 4.

In 1946 George Gamow (1904 – 1968) took Hubble's law to its extreme and argued that, since the universe has been expanding all this time, it must have been very small at some point in the past. Using the known laws of physics, he argued that in the early universe the matter density must have been so high that rapid thermonuclear reactions had to be going on. This lead to the Big Bang Theory of the creation of the universe. Following this, in 1948 R. A. Alpher (1921 –), H. A. Bethe, and G. Gamow predicted that the blackbody radiation from the original Big Bang should still fill the universe. Originally Gamow had inserted Bethe's name (with "in absentia" in brackets) as a joke, to make the names like the first three letters of the Greek alphabet: alpha, beta, gamma. Bethe happened to be a reviewer. He added some important details and, as a further joke, crossed out the (in absentia).

Robert Henry Dicke (1916 – 1997) came up with the same idea in 1964. This led to the following series of events. In 1965 Arno Allan Penzias (1933 –) and Robert Woodrow Wilson (1936 –) observed some radiation noise in an antenna they had set up. This radiation is now known as the residual black body radiation from the Big Bang.

Michael Turner, a well-known cosmologist at the University of Chicago evaluated their work as follows. "The discovery of the cosmic microwave background by Penzias and Wilson transformed cosmology from being the realm of a handful of astronomers to a 'respectable' branch of physics almost overnight." John Bacall, an astrophysicist at the Princeton Institute for Advanced Studies gave no less of an evaluation. "The discovery of the cosmic microwave background radiation changed forever the nature of cosmology, from a subject that had many elements in common with theology to a fantastically exciting empirical study of the origins and evolution of the things that populate the physical universe." Here was proof that the universe was born at a definite moment, some 15 billion years ago.

Penzias colleague, Ivan Kaminow, joked that Penzias was an unusually lucky guy. "Arno Penzias and Bob Wilson were trying to find the source of excess noise in their antenna, where pigeons were roosting. They spent hours searching for and removing the pigeon dung. Still the noise remained, and was later identified with the Big Bang. Thus, they looked for dung but found gold, which is just opposite of the experience of most of us."

At first, Penzias and Wilson were at a loss to explain their result until Philip James Edwin (Jim) Peebles (1935 –), who was working in a group directed by Robert Dicke at Princeton, pointed Penzias to a 1964 paper by Dicke. After discussions with Dicke and Peebles, Penzias and Wilson published a paper detailing the experimental results. This paper was accompanied by a theoretical paper by Peebles and Dicke. Some time later Penzias discovered the earlier Gamow paper that had already contained the theoretical results. After sending a copy of their own papers to Gamow he received a reply which outlined all the work that Gamow had done on the subject and ended with the sentence, "So you see that the world did not start with almighty Dicke."

Gamow apparently was usually more interested in the answer than in the rigor of the method by which it was obtained. Bethe estimated that Gamow's popular books, of which there were several, were probably 90% correct. Teller used this as an argument in favour of Gamow's writings by stating that a book which is 99.44% (the advertised purity of ivory soap) correct would prove exceedingly dull.

According to William A. Fowler (1911 – 1995), Gamow, did a lot of work on cosmological time scales and accordingly had an exceedingly accurate sense of the value of time. As an example he cited the case where at a very dull meeting in New York, Gamow turned to him with the words, "Sonny Boy, why are we wasting our time here? Let's go have a drink." This they did.

Fowler's main work was in astrophysics and nucleosynthesis in massive stars and supernovae. His research led him to conclude that almost all heavy atoms were formed in stars. He phrased this as follows, "All of us are truly and literally a little bit of stardust." Much of this work he carried out at Caltech at the Kellogg Radiation Laboratory named after the magnate of breakfast cereal fame. When an elderly woman found out where he worked and asked him what he did there he tried to explain, but his explanation was far too technical. She came back with, "Young man, I still don't know what you do." So, he tried a less technical approach, "I bust atoms and study how things explode." This time she understood, "Oh, you puff rice."

In 1948 Sir Herman Bondi (1919 –), Thomas Gold (1920 –), and Fred Hoyle (1915 – 2001) proposed an alternative to the Big Bang, the "steady-state cosmology" in which matter is continuously created. The rate was exceedingly small, about ten hydrogen atoms per cubic meter per million years. In this theory the universe has no beginning and no end with average properties that remain constant throughout space and time. Bondi also stated, "Indeed one can argue that science is only possible because one can say something without knowing everything."

Until the discovery of the cosmic blackbody radiation, there was a running feud between the proponents of the Big Bang Theory: Alpher, Gamow, and Herman and the proponents of the Steady-State-Cosmology: Bondi, Gold, and Hoyle. It was Bondi who first referred to the theory of Alpher, Gamow and Herman, with some sarcasm, as the "Hot Big Bang". The name stuck. When Alpher and Herman found an error in one of Hoyle's papers they wrote him a detailed letter so that he could respond before they published a review paper. They never received an answer. Willy Fowler gave the reason for this. At the time, Hoyle had a very popular BBC lecture series and received on the order of 30, 000 crank letters. Fowler suggested that the letter by Alpher and Herman received the same fate as all those other letters. Neither side ever convinced the other.

This, and the fact that in the forties cosmologists were happy if their theories agreed with observations to within factors of ten or even a hundred may be what led Landau to say, "Cosmologists are seldom right, but never in doubt."

Incidentally Fred Hoyle also wrote science fiction with great success, but his views on space travel were unambiguous. "The Soviet-American space race is almost worthless for scientific research. What has been accomplished is not worth a thousandth part of what has been spent."

Max Born expressed a similar sentiment several times. Thus, in 1958 at a conference to the Evangelical Academy in Kloster Locain he pronounced, "Space travel is a triumph of intellect but a tragic failure of reason. [It is] a symbol of a contest between the great powers, a weapon in the cold war, an emblem of national vanity, a demonstration of power." He repeated this theme in 1961 on the TV Program *The Voyage into the Dark*, "Intellect distinguishes between the possible and the impossible; reason distinguishes between the sensible and the senseless. Even the possible can be senseless."

John Cockroft also disparaged manned space flight. "We smile as we watch your space flights on television. Your efforts represent a distortion of science in the name of competition with the Soviet Union."

In 1966 Edward Condon (1902 – 1974) was asked to head a study of UFO's for the US air force. Since Condon pulled no punches his study led to some rows. A reporter who interviewed Condon at that time remarked, "Dr. Condon you seem to have been the center of controversy fairly frequently."

A smiling Condon replied, "It is not very hard."

One of the early workers on theoretical aspects of the thermal properties of empty space, the vacuum, was the Soviet physicist, D. A. Kirzhnitz. When he finished his Ph.D., the Soviet government sent him to work in a factory. This was standard procedure at the time and as a young Soviet scientist he had no choice but to go. At the factory the director asked him what he had worked on before he came. He answered, "The vacuum". The director was delighted, "Here is an instrument, go and find the vacuum leaks in those pipes."

The modern guru of cosmology is Stephen W. Hawking (1942 –). Don Page was visiting the Hawkings when Mrs. Hawking received a telephone call from Dirac saying that he wanted to talk to Stephen Hawking. Don Page explained to Mrs. Hawking who Dirac was. At this time Stephen Hawking was still able to talk but with great difficulty and not too understandably. Dirac, of course, was famous for his terseness. Thus, the two famous physicists sat across from each other barely speaking a word. Only the wives carried on a lively conversation.

To break the painful silence when the wives left the room Don Page asked Dirac what he was working on. The answer was barely audible and Page did not catch the answer. However, after a further silence Dirac turned to Page and asked, "Did you know that Olivia Newton-John is Born's granddaughter?"

On a TV program about Hawking the Oxford relativist, P. K. Tod, began a lecture at Cambridge. He wrote the metric, $g_{\mu\nu}$ on the board. At this point, Stephen Hawking asked, "What metric are you using, Lorentzian or Euclidean?"

"I'm from Oxford, so of course I'm using the Lorentz metric."

"Oh! We gave that up years ago!" replied Hawking.

Hawking has also been the champion of the Many-worlds Interpretation of quantum mechanics. The same was true of another famous cosmologist, Bryce Seligman DeWitt (1923 –).

The reason for this peference is that modern cosmologists even attempt to write a wavefunction for the whole universe. The physical interpretation of such a wavefunction is meaningless at worst or obscure at best if one tries to use the Copenhagen interpretation since there is no "observer" to record an observation. The Many Worlds Interpretation can avoid this difficulty by assuming that every time an event occurs with more than one possible outcome, all of these outcomes actually happen: each in a different universe. This seems to avoid the necessity for an observer.

Ray F. Streater (1936 –) on his homepage lists the Many Worlds Interpretation among "Lost Causes in Theoretical Physics" and says that "This subject arose in cosmology; quantum mechanics was introduced in the belief that it must be used in any good model of the universe as a whole. This is misguided: the large scale dynamics of galaxies is just the sort of domain in which the delicate phase relations between components of a quantum state become washed out, and classical mechanics takes over." With regard to the Many Worlds Interpretation he states: "There is nothing to the many-worlds theory. There are no theorems, conjectures, experimental predictions or results of any sorts."

Still, it must be conceded that the Many Worlds Interpretation is much favoured by science fiction writers who have produced interesting stories.

When Bryce DeWitt ran into Dirac at a conference, he took the opportunity to explain his work to him. Dirac listened patiently, nodding his head occasionally. After some time DeWitt paused and asked Dirac if he had any questions. Dirac answered, "Yes, where is the men's washroom?"

Dirac was a frequent visitor to the Soviet Union. In 1954 The US government embarrassed its physics community by refusing a visa for a visit that Dirac had planned to Princeton. As a consequence the Canadian physicist, Gerhard Herzberg (1904 – 1999), invited Dirac to a symposium in his honour in Ottawa, a symposium which his American colleagues could attend. Herzberg had achieved recognition when he was the first to detect molecules in outer space. He received the 1971 Nobel Prize in Chemistry "for his contribution to the knowledge of electronic structure and geometry of molecules, particularly free radicals." About his work he said, "I don't have many problems which are brilliant, but if I have a problem that I think is important, I persist in it — in a sense I am a beaver." The beaver is one of the symbols of Canada.

When the cosmologist, Vladimir Fock (1893 – 1974) was invited to give a talk at the Lebedev Institute he asked his hosts for the promised honorarium, before the talk. They said they would settle all this after the talk. Fock insisted that he would not present his talk until he had the money in his hands. In 1948, as quoted by M. Nussenzweig, he also stated the case for simplicity, "The aim of a theory is to give a picture reproducing all the qualitative and quantitative features of the phenomenon considered. This aim is not attained until the solution obtained is of a sufficiently simple form."

Throughout the rest of his life Einstein tried to unify gravity and electromagnetic theory. The same was true of Schrödinger. In 1943 Schrödinger thought he had succeeded and considered it as great as anything he had ever done. "I have found the unitary field equations. They are based only on primitive affine geometry, a way which Weyl opened and Eddington extended, whereupon Albert did the main job in 1923, but missed the goal by a hair's breadth. The result is fascinatingly beautiful. I could not sleep for a fortnight without dreaming of it."

Einstein, after all his failures at unification, was less enthusiastic. "Concerning an affine solution of the electrical problem, I have become quite skeptical. This thing alongside so many others, has been relegated to a pretty spot in the graveyard of my enthusiastic hopes — at the time I found it dif-

ficult to separate myself from it. . . . One thing is certain. The Lord has not
made it easy for us. As long as one is young, one does not notice this much
— luckily."

Regarding the unification of all known forces one might want to keep in
mind the following comments. The first one is by Freeman Dyson. "The
history of physics is littered with the corpses of dead unified theories." Pauli
also had this to say with respect to Einstein's attempts to unify gravity and
electromagnetism, They "cannot be joined for God hath rent them asunder."

Similarly, Richard Feynman stated in 1962, "None of these unified field
theories has been successful. . . . Most of them are mathematical games, invented by mathematically minded people who had very little knowledge of
physics and most of them are not understandable."

According to A. Trautman, "A principle is a statement about physical
theories extrapolated from experiments and experience. Example: Principle
of Equivalence, gravitational and inertial mass are the same."

There is one more cosmological theory that was invented by Dirac in 1937,
shortly after he got married. He noted that the ratio of the electric force to
the gravitational force of attraction of the electron in an Hydrogen atom is
equal to the ratio of the age of the universe to the typical nuclear lifetime.
This would not have been of interest, except that these ratios are both an
incredibly large number: 10^{39}. Now such large numbers do not arise from
normal computations and Dirac argued that these numbers must always have
been equal so that in the distant past, both numbers were close to 1. This
meant that these large ratios could only be explained if the gravitational
constant decreased with the age of the universe.

After Dirac published this idea, Bohr one day rushed into his institute,
with a copy of *Nature* containing the article, in his hand, and shouted, "Look
what happens to people when they get married."

The idea did not catch on and Dirac abandoned it for almost thirty years
when he once again returned to it. In the meantime the same Pascual Jordan
that had been instrumental in the development of matrix mechanics had
also taken up this idea. He tried to make the gravitational constant into a
scalar field in its own right and published his version in a book *Schwerkraft
und Weltall* (Gravity and the Universe). The publisher had a problem with
the first copies and some of them contained only blank pages. When Pauli
received one of these blank copies he is supposed to have said, "Jordan knows
that I can think up what should be in it by myself." Pauli also later continued
to encourage Jordan to work on this theory. Even so, Pauli could not resist

being Pauli. After Jordan had presented a talk on his theory, Pauli who had been sitting in the front row shaking his head from side to side in his usual fashion, rose to speak, "A theory, Herr Jordan, does not become true just by talking about it." At present this theory, along with numerous other unified theories, lies ignored and forgotten.

Pauli's friendship with Jordan is all the more astonishing since the latter had joined the Nazi party and even worn the uniform of a storm trooper. Still, even during the darkest days of the Nazi regime, he championed relativity and did not hesitate to use Einstein's name. He even ridiculed the attempts to define German science by stating, "The differences between German and French mathematics are no more than the differences between German and French machine guns." Still he did write some articles reeking of Nazi ideology. When Pauli once quoted some of these writings back to him and asked, "Herr Jordan, how could you write such a thing?" Jordan quickly responded, "Herr Pauli, how could you read such a thing?"

After the war Pauli helped to get Jordan rehabilitated. "It would be incorrect for West Germany to ignore a person like Pascual Jordan." After that, Jordan was promoted from visiting to full professor at the University of Hamburg.

Chapter 10

Black Holes

"Abandon all hope, ye who enter here." Dante Alighieri, *The Inferno.*

For most people, the concept of a black hole is associated with Einstein's General Theory of Relativity. Actually the idea of a black hole is much older than that and is already contained in Newton's theory of gravity. A black hole is nothing more than an object so massive that, due to its gravitational attraction, nothing can escape from its surface, including light. As early as 1783 John Michel (1724 – 1793) had presenteda paper (via Faraday) to the *Royal Society* that "the attractive force of a heavenly body could be so large that light could not flow out of it." Michel was a teacher at Cambridge for fourteen years and later the rector of Thornhill in Yorkshire. He was led to his ideas by Newton's theory of gravity as well as Newton's *Opticks* in which that great philosopher of science had proposed that light consisted of particles.

Michel not only proposed the existence of these invisible stellar bodies, but went on to suggest means by which they might be detected. Somehow his work was forgotten soon after his death.

Fifteen years after Michel's proposal, in 1798, Pierre Simon de Laplace (1749 – 1827) wrote, "A luminous star, of the same density as the Earth, and whose diameter should be two-hundred and fifty times larger than that of the sun, would not, in consequence of its attraction, allow any of its rays to arrive at us. It is therefore possible that the largest luminous bodies in the universe may, through this cause be invisible." Michel's work is now usually forgotten and Laplace is credited as being the first to propose the existence of what are now called "black holes".

From the gravitational acceleration of an object on its surface and from its size it is possible to calculate the escape velocity. This is the minimum speed with which an object has to be projected away from the surface of that body and not fall back. For the earth this speed is about 11 km/s and is the minimum speed required to put satellites into orbit. For the sun this speed is 618 km/s. For a black hole the escape velocity has to exceed the speed of light, about 300,000 km/s. So, for the earth to be a black hole would require that all the earth's mass be confined to a sphere with a radius of about 8.9 mm. Similarly, the mass of the sun would have to be inside a sphere with a radius of about 3.0 km. For this to happen, the density of matter would have to be very large. However, for a very large object, like the one described by Laplace, the density of matter would be like that of the earth, and for even much larger objects the density could be that of air.

It was one thing to propose the possibility that black holes might exist, it was a totally different matter to see how such objects could actually come to be. The idea of objects as large as several hundred times the size of the sun or objects with a density 100,000 times as great as that of the sun seemed too outrageous to take seriously. Before such ideas could be considered, more than two centuries passed.

What is it about black holes that has gripped the popular imagination? Why has this idea given rise to numerous science fiction stories and even a movie by the same name? Perhaps because one can never, even in principle, enter a black hole and send back information about what is inside. Not only that, but a black hole will continue to attract more matter from around it and continue to grow. Thus, it is a sort of cosmic vacuum cleaner. Why did John Wheeler name it a black hole? The answer is simple. It absorbs all light falling on it and is thus absolutely dark. It can reveal its existence by having a visible companion that orbits around it. Otherwise when we look into the cosmos, a black hole presents exactly that a circular region of blackness.

There is another way a black hole can reveal itself. It can act as a strong lens. Because of its gravitational pull, light from a star directly behind the black hole would be bent and therfore focussed as it passed the black hole.

The modern theory of black holes hails back to the same astronomer, Karl Schwarzschild (1873 – 1916), who had asked Max Born (during his oral examination) what observations he would make on a falling star. During World War One, Schwarzschild volunteered for service in the German army. While serving on the Russian front in 1916, he produced the first two exact solutions of the Einstein field equations. Until then, Einstein had used only approximate solutions for his predictions. One of Schwarzschild's solutions corresponds to the metric for empty space outside a massive body such as a black hole and is called the "Schwarzschild metric".

The Schwarzschild metric has a singularity (becomes mathematically infinite) at what is called the Schwarzschild radius which is determined by the mass of the object. In 1922 at a Paris conference, the mathematician Jacques Hadamard (1865 – 1963) asked what was the significance of this singularity. Would it ever be possible for a physical system to attain this singularity? Albert Einstein insisted that this could not happen. He pointed out the dire consequences for the universe, and jokingly referred to the singularity as the "Hadamard disaster".

For black holes to be accepted required a mechanism by which an ordinary star could turn into a black hole. The natural process was for a star to collapse due to its own gravitational field, so-called "stellar collapse". As a star burns its nuclear fuel, the pressure from its radiation and its gravitational pull may balance. This can go on until the fuel is exhausted. After that it is only the Pauli Exclusion Principle that can prevent collapse.

The idea of black holes started to be accepted as a serious scientific concept after a number of astronomical observations pointed to the existence of very dense matter. The first such indications came from two dim stars: the companion stars of Sirius and Procyon. The puzzling thing about these companion stars was that, although they emitted white light like their bright partners, they were a hundred times dimmer. By now astronomers knew that the light emitted by any luminous body, such as a star, was that of a blackbody radiator. The spectrum and therefore the brightness were completely determined by the surface area and the temperature of the object. So, how could a star radiating white light be so dim? One proposal was that these companions were simply reflecting the light emitted by their bright partners. Another, less acceptable, proposal was that these dim stars were of much greater density and so much smaller surface area.

In 1927 Sir Arthur Eddington phrased this as follows. "The message of the companion of Sirius when it was decoded ran: 'I am composed of material 3000 times denser than anything you have ever come across; a ton of my material would be a little nugget that you could put in a matchbox.' What reply can one make to such a message? The reply which most of us made in 1914 was — 'Shut up. don't talk nonsense.' "

The companions of Sirius and Procyon were the first objects to be identified as "White Dwarfs". What was needed was a mechanism to explain the existence of these objects, a mechanism to explain how ordinary matter could become so extremely dense.

Eddington then (1925) proposed that the outer electrons in the Bohr atoms might be stripped off if the star were sufficiently hot to ionize its atoms. This would make the atoms constituting the star smaller and allow for much denser material. This mechanism was fraught with another theoretical difficulty also proposed by Eddington. As the star cooled, its atoms would again regain their electrons and so return to their original size. However, iif they did that, the whole star would have to expand against the force of gravity. This would require energy that a cold star did not have. The proposed mechanism semed self-contradictory.

The answer to the puzzle posed by white dwarfs started to emerge in 1930 from the work of a nineteen year old on board a ship from India to England. He found that using relativity theory and quantum mechanics it was possible for a sufficiently massive star to have gravity overcome the Pauli repulsion. Subramanyan Chandrasekhar (1910 – 1995) also showed that there is a limit — the Chandrasekhar limit — to the largest mass a white dwarf star can attain.

Later, Chandrasekhar refined his computations and showed that stellar collapse was definitely possible. Eddington viciously attacked the theory underlying this result. Stellar collapse could not occur. He argued that in a "correct" calculation, special relativity had to be abandoned in favour of Newtonian mechnics. His arguments wereused nonsensical and contradictory, but they delayed the acceptance of black holes for several decades.

In a letter home, Chandrasekhar wrote, "The differences are of a 'political' nature. Prejudices! Prejudices! Eddington is simply stuck up! Take this piece of insolence. 'If worse comes to the worst we can believe your theory. You see I am looking at it from the point of view not of the stars but of Nature.' As if the two are different. 'Nature' simply means Eddington personified as an Angel! What arguments could anyone muster against such brazen presumptuousness?" Yet, despite their scientific animosity, the two somehow remained friendly on a social level.

Subramanyan Chandrasekhar won the 1983 Nobel Prize in physics "for his theoretical studies of the physical processes of importance to the structure and evolution of the stars." He shared this prize with William A. Fowler who received it "for his theoretical and experimental studies of the nuclear reactions of importance in the formation of the chemical elements in the universe."

In 1937 Chandrasekhar moved to Chicago where he remained. So, when in 1951 he visted Bangalore, Raman noted that he had not been back for a long time and told him, "You are visiting India like Halley's comet."

Although white dwarfs could be extremely dense, they were limited in size and thus could not collapse into black holes. When Chadwick discovered the neutron, several physicists immediately considered the possibility of stars made of nothing but neutrons: so-called neutron stars. In 1939 J. Robert Oppenheimer and George Volkoff proved that neutron cores like white dwarfs were limited in how heavy they could be. If the neutron core exceeded 70% of the sun's mass its state was unstable. Any core larger than this had to continue to collapse, but to what?

In 1939 Robert Oppenheimer and his student Hartland Snyder (1913 – 1962) provided the answer to the question above in a paper in *Physical Review 56, 455, (1939)*. They wrote, "When all thermonuclear sources of energy are exhausted a sufficiently heavy star will collapse. Unless fission due to rotation, the radiation of mass, or the blowing off of mass by radiation, reduce the star's mass to the order of that of the sun, this contraction will continue indefinitely."

They concluded that, "The star thus tends to close itself off from any communication with a distant observer; only its gravitational field persists."

Here, for the first time, was a definite proposal for how a physical object could become a black hole.

Only a few months earlier, Einstein had published a paper in which he argued that black holes could not arise due to the gravitational collapse of a star. His conclusion was due, not to faulty mathematics, but to unrealistic assumptions.

In 1954 Maurice Goldhaber (1911 –) got into into an argument with his host, Hartland Snyder, over the existence of the antiproton. The Dirac equation predicted the existence of just such a particle. Goldhaber maintained, "I said, 'I don't believe it until it is proven.' " He pointed out, there was the inconvenient detail that the universe around us is made of matter, not antimatter. Snyder reached out, grabbed his hand, and said, "I bet you $500 that the antiproton exists." Goldhaber accepted the bet without thinking even though his brother Gerson Goldhaber was at that moment trying to find antiprotons at Lawrence Berkeley Laboratory.

Goldhaber lost the bet; the antiproton was discovered in 1955 by Emilio Segrè and Owen Chamberlain. For this they received the 1959 Nobel Prize in physics.

Why, if none were detected, were physicists interested in black holes and why did they write learned tomes about this object? There are two reasons. One is that a black hole is, in a sense, the simplest of all possible objects. This was realized early on when Werner Israel (1931 –) proved that any static black hole is spherical and completely described by its mass and charge. this was the first of the so-called "No-hair Theorems". He also coined the term "event horizon" for the spherical region around a black hole from which nothing escapes. A few years later, in 1970, Brendan Carter showed a result whose consequence was that all black holes are axially symmetric and completely described by three numbers: their mass, charge, and angular momentum. This led John Archibald Wheeler (1911 –) to call these results the *No-hair Theorem*: "Black holes have no hair." In other words, they have no structural features.

Werner Israel and his wife Inge arrived in Edmonton in 1958. At that time Edmonton was still rather provincial and had very few delicatessens. One day Inge sent Werner downtown to get a barbecued chicken. He duly took the bus and bought the chicken. However he did not return home with it until two hours past dinner time. As he had walked along the main street he noticed that a particular movie, that he had wanted to see, was playing. So, he had stopped in and watched the movie before bringing the chicken home.

On another occasion Inge sent him to buy some pastries since guests were expected for an afternoon tea. When he returned home, half the pastries had disappeared; he had sampled "a few" on the way home. According to Inge — a statement Werner vehemently denies — there were not enough pastries left for the guests.

One year Werner Israel was given an undergraduate course in electromagnetic theory to teach to engineers. Now, like most theorists, he always worked in Gaussian units whereas engineers always work in MKS rationalized units. His friend Harry Schiff asked him if he found it painful to learn the MKS rationalized units. His reply was, "I don't have to learn these units, I only have to teach them."

Regarding the use of different units, Feynman had this to say in *The Character of Physical Law*. "For those who want some proof that physicists are human, the proof is in the idiocy of all the different units which they use for measuring energy."

On an invited visit to Vienna, Israel and his wife Inge were riding a tram. As it approached an intersection where it would make a sharp turn, the passengers anticipating this reached up to grasp the straps hanging there for that purpose. Israel did the same, only his strap was bright red. As the train swerved around the corner it came to a sudden stop. The centripetal acceleration had pulled Israel to the side and he had pulled the emergency brake. According to him the Viennese politely tried not to look at him as he went to the front to explain to the conductor who was desperately pulling levers to get the tram out of the intersection where it blocked rush hour traffic for about ten minutes. Inge explained the reason for the conductor's lack of interest in Werner. "We were only a block or so away from the Boltzmann Institute and the conductor must have realized that this was just another physicist."

In April, 2003 Israel gave a talk, reviewing the known results regarding the Beckenstein-Hawking formula for the entropy of a black hole. During the question period someone asked whether this entropy had anything to say about the inside of the black hole. More precisely, did he not agree that "re-radiation at the event horizon of the blackhole would keep information from getting lost?" Israel answered, "This is a beautiful hand-waving argument. I would like to believe it, but I find it hard."

Werner Israel told me the following story about his fellow relativist H. P. Robertson (probably the referee that rejected Einstein's paper). In his early years as a physicist Robertson spent some time in Germany and even learned the language. But on one occasion he confused the two verbs, *wiegen* (to weigh) and *wagen* (to dare). He was in a hurry and rushed into a post office and asked the young lady behind the counter, *"Haben Sie eine Wiege ich möchte Etwas wagen."* ("Do you have a cradle I would like to dare something.") instead of the intended *"Haben Sie eine Waage ich möchte Etwas wiegen."* ("Do you have a scale I would like to weigh something.")

After the results by Israel and their extension by Brendan Carter, Stephen W. Hawking and G. F. R. Ellis, in 1973, published a book of rigorous mathematical theorems on the structure of space-time. One of the results they obtained was that any singularity in the space-time structure would be hidden behind an event horizon. This was called the "Cosmic Censorship Hypothesis" and is usually worded, "There are no naked singularities." This hypothesis was first stated by Sir Roger Penrose(1931 –).

Hawking and Ellis' proof that physical singularities could occur only inside black holes where they cannot be seen, was challenged by Kip Stephen Thorne (1940 –) and John Preskill (1953 –). They believed that (naked) singularities could exist. In 1991 Hawking bet with Thorne and Preskill that such singularities could not exist. However, after supercomputer simulations showed how a naked singularity could exist, Hawking was forced to concede the bet (according to Hawking "on a technicality") on Feb. 5, 1997. In conceding, Hawking presented his colleagues with "adequate raiments to shield their nakedness from the vulgar view."

In the early 1970's Denmark was the most sexually liberated country in the world. Strip shows abounded throughout Copenhagen. This led Willy Fowler to claim to have experimental proof that naked singularities do not exist. To celebrate Fowler's sixtieth birthday in 1971 there was a conference in Denmark and it is reported that during his stay in Denmark, Sir Roger Penroselooked very hard for naked singularities and did not find any. According to Fowler, this proved that naked singularities don't exist since, at that time, if there was anything naked to be found, it was in Denmark.

Only a year earlier Penrose had this to say about space-time singularities. "When eventually we have a better theory of nature, then perhaps we can try our hands, again, at understanding the extraordinary physics which must take place at a space-time singularity."

One day after conceding the bet to Preskill and Thorne, Hawking made another bet; this time Thorne sided with him. This bet was worded as follows.

"Whereas Stephen Hawking and Kip Thorne firmly believe that information swallowed by a black hole is forever hidden from the outside universe, and can never be revealed even as the black hole evaporates and completely disappears,

"And whereas John Preskill firmly believes that a mechanism for the information to be released by the evaporating black hole must and will be found in the correct theory of quantum gravity,

"Therefore Preskill offers and Hawking/Thorne accept, a wager that:

"When an initial pure quantum state undergoes gravitational collapse to form a black hole, the final state at the end of black hole evaporation will always be a pure quantum state.

"The loser(s) will reward the winner(s) with an encyclopaedia of the winner's choice, from which information can be recovered at will."

The document was signed by Thorne and Preskill with a thumbprint by Hawking and was dated Feb. 6, 1997 at Pasadena California.

At GR17 (The 17th International Conference on General Relativity and Gravitation) in July 2004, held in Dublin, Ireland and in front of a large

audience of physicists and reporters, Hawking conceded. He first presented
a talk explaining the calculations he had performed that led him to concede.
At the end he stated, "In 1997, Kip Thorne and I bet John Preskill that
information was lost in black holes. The loser or losers of the bet are to
provide the winner or winners with an encyclopaedia of their own choice,
from which information can be recovered with ease. I'm now ready to concede
the bet, but Kip Thorne isn't convinced just yet. I will give John Preskill
the encyclopaedia he has requested. John is all-American, so naturally he
wants an encyclopaedia of baseball. I had great difficulty in finding one over
here, so I offered him an encyclopaedia of cricket, as an alternative, but
John wouldn't be persuaded of the superiority of cricket. Fortunately, my
assistant, Andrew Dunn, persuaded the publishers Sportclassics Books to fly
a copy of *Total Baseball: The Ultimate Baseball Encyclopedia* to Dublin. I
will give John the encyclopaedia now. If Kip agrees to concede the bet later,
he can pay me back."

Preskill accepted the encyclopaedia and then waving this trophy over his
head said, "I always hoped that when Stephen conceded, there would be a
witness — this really exceeds my expectations." He also admitted that "I'll
be honest. I didn't understand the talk." He then went on to explain that
he would need to see more details. Thorne's reason for not yet conceding
was in a similar vein. He also said that he would need to study Hawking's
argument more closely.

I finish with another recent comment on black holes by Martinus Veltman who shared the 1999 Nobel prize in physics with Gerardus 't Hooft
(1946 –) "for elucidating the quantum structure of electroweak interactions
in physics." (1931 –). In 2001 at the Heisenberg Centennial in Bamberg,
Veltman expressed his opinion as to why Einstein did not accept quantum
mechanics. "Very likely Einstein felt all along that General Relativity and
Quantum Mechanics are not compatible. I always get nervous about black
holes."

Chapter 11

Epilogue

"Brief is this existence, as a fleeting visit in a strange house. The path to be pursued is poorly lit by a flickering consciousness, the center of which is the limiting and separating I." Albert Einstein 1954.

Most of the developments, that I have sketched, stop somewhere in the 1980's. This is not because there haven't been any major developments in physics since then. Indeed there has been major progress in condensed matter physics, especially in the manipulation of individual electrons, something that has led to the rapidly growing field of nanotechnology. It has now been possible for some time to obtain pictures of individual electrons in an atom. Of course the pictures are not taken with optical instruments, but the images we get are precisely those that have been illustrated in physics texts from computations with the Schrödinger equation.

In the realm of cosmology there has also been major progress in the experimental verification (and falsification) of certain models. Dark matter and dark energy seem more a reality now then twenty years ago. Also, when I was a student, black holes were still very speculative objects more likely to exist in the imagination of theoretical physicists than in the cosmos. Today we have convincing evidence of black holes. Not only that, strong evidence exists that at the center of our own Milky Way galaxy there is a gigantic black hole.

Yet, many questions remain at a very elementary level. What determines the value of the fine structure constant? Why are there only three perceived spatial dimensions? Is renormalizability of a theory a physical or simply computational equirement?

There is also the problem (in spite of claims by practitioners of String Theory) of how to reconcile General Relativity and Quantum Mechanics. Norbert Wiener phrased this as follows. "The modern physicist is a quantum theorist on Monday, Wednesday, and Friday and a student of gravitational relativity theory on Tuesday, Thursday, and Saturday. On Sunday he is neither, but is praying to his God that someone, preferably himself, will find the reconciliation between the two views."

There is one more aspect of physics that, although displayed in the stories, has not been emphasized. Physics is international. Of all fields of human endeavour it is the most international. This already led Pauli to comment half a century ago. *"Die Physik (und sowie Physiker) ist eine ambulante Wissenschaft. Physiker sind zuerst Physiker und dann erst Deutsch, Französich, usw."* ("Physics as (well as physicists) is an ambulant science. Physicists are physicists first and German, French, etc. after that.")

Physics is not a cold synthesis of objective experimental data. There is no such thing as a "scientific method". The discoveries of physicists are greatly influenced by their individual, preconceived, subjective ideas and prejudices. Intuition, luck, and sheer audacious guess work go into the creation of physical theories. At more than one physics conference have I seen someone in the audience stand up and shout at the speaker, "You are an idiot," when their favourite belief was put in doubt. Physicists are far from dispassionate in describing the physical world. The main difference between science and other human endeavours is not that scientists are more objective, rather that their ideas can and are put to the test. This is not always sufficient to convince everyone, since having invested so much of their life into furthering a given idea they resist giving it up. It is as Planck said, "An important scientific innovation rarely makes its way by gradually winning over and converting its opponents: it rarely happens that Saul becomes Paul. What does happen is that its opponents gradually die out, and that the growing generation is familiarized with the ideas from the beginning."

In a similar vein Bohr said, "We are no wiser and no less biased than other people. But as a physicist, or a biologist, you are certain to have gone through the experience of making a confident assertion, and then being proved wrong. A philosopher or a sociologist might never have had this wholesome lesson."

There is one more point regarding science. Science does not provide or even claim to provide any absolute truths. It is as Niels Bohr said, "It is wrong to think that the task of physics is to find out how nature is. Physics concerns what we can say about nature." In this regard science differs from religion. Good science does, however, state the limits of its validity. Scientists are forever trying to push the limits of a theory to see where it breaks down. This breaking down, does not invalidate the theory. It delimits the region of applicability of the theory; it shows how far the theory may be pushed. Newton's theory is as valid today as it was three-hundred years ago. It simply does not apply to extremely small, extremely fast, or extremely massive objects. It is precisely a recognition of this limitation of all theories that makes science so useful and powerful.

It is the desire of physicists to push the limits of validity of a theory that prompted Y. Takahash to tell me years ago that we don't solve problems, we create new problems. He phrased this as a metaphor about someone who has one sleeve too short and pulls on it only to find the other too short. This inspired me to write the following poem.

Problems

The right sleeve is short; the left sleeve is right.
So tug on your right sleeve with all your might.
The sleeves' situation is now in reverse.
The normal condition is mostly perverse.

In one respect, science is very much like religion: science requires faith. Sir Humphrey Davy stated this most eloquently in his *Consolations in Travel — Dialogue V — The Chemical Philosopher*. "Scientists were rated as great heretics by the church, but they were truly religious men because of their faith in the orderliness of the universe."

When individuals claim to have knowledge of an absolute truth, they are stating that no meaningful questions regarding this "truth" may be posed. Galileo Galilei gave the response to this more than four centuries ago when he stated, "I do not feel obliged to believe that the same God who has endowed us with sense, reason, and intellect has intended us to forgo their use."

So, why stop in the 1980's? The reasons are two-fold. The newest developments are not yet in the same category as the great developments of the first half of the twentieth century. Maybe in a decade or two we will know which of these developments will change our view of the world as dramatically as those that I have described. In my opinion, big surprises still await us. Will they come from string theory or some of as yet embryonic ideas? Will quantum mechanics have to be modified to make it consonant with relativity or will relativity have to be modified or will some totally new approach be required? Or has string theory already achieved this? Will the application of physics to biological systems be one of the major new developments or will outrageous ideas about the origin of the universe dominate? Will physics subdivide even further or will there be a new fusion of this discipline?

At the beginning of the second half of the twentieth century it was still possible for a physicist to make major contributions in all areas of physics. Today this seems impossible; the field is too diverse. This led Wigner to lament some time ago. "Physics is becoming so unbelievably complex that it is taking longer and longer to train a physicist. It is taking so long, in fact, to train a physicist to the place where he understands the nature of physical problems that he is already too old to solve them."

The second reason for stopping in the 1980's requires some explanation as to why I wrote this book as well as *Quips, Quotes, and Quanta*. My generation was pretty well the last one that got to meet some of the founders of what we call "modern physics". One or two of my graduate students may have actually seen Richard Feynman. They have very likely seen or even met Murray Gell-Mann, but even to them, most of the names from *Quips, Quotes, and Quanta* are simply names associated with ideas that they take for granted; they are not names associated with real people. On the other hand I saw how, in my lectures, students responded to the anecdotes I told about these creators of the ideas they had to study. It made these ideas more alive. That is why I felt it was necessary to preserve these stories. I hope that somewhere, in the physics community, are young physicists noting down little stories of their heroes of today so that they can also pass them on to future generations.

Chapter 12

Glossary and Timeline

12.1 Glossary

big bang theory holds that the universe was created about 10 to 20 billion years ago in a gigantic explosion of a highly concentrated mass of matter.

boson a particle with integer spin.

cosmic background radiation refers to the residual blackbody radiation from the big bang.

cosmic rays are a steam of high energy charged particles throughout space.

cosmological constant was introduced by Einstein into his field equations to obtain a static unchanging universe.

critical mass is the minimum mass at which a fissionable material such as uranium 235 will produce a self-sustaining chain reaction.

cyclotron is a particle accelerator consisting of two D-shaped pole pieces—dees—separated by a gap with a magnetic field perpendicular to the dees. An alternating electric field across the dees acclerates charged particles injected into the gap.

Dirac sea refers to the completely filled negative energy states arising from solutions of the Dirac equation.

fermion a particle with half odd integer spin.

hadron is any particle which interacts strongly.

Hall effect an electric field in a conductor. It is perpendicular to both the current through the conductor and a magnetic field applied externally to the conductor

holes are the empty states in the Dirac sea.

Hubble constant is the proportionality constant between the rate at which galaxies are receding and their distance. Its value is between 50 and 100 km/sec per megaparsec.

isospin is a quantum number that allows neutrons and protons to be treated as two states of the same particle, but with different isospin.

Josephson effect describes the flow of superconducting electron pairs across a thin dielectric barrier separating two superconductors.

Lamb shift is a small energy difference in two levels of the hydrogen atom that would be the same according to Dirac theory except for some corrections due to quantum electrodynamics.

magnetic monopoles are hypothetical particles that carry either a north magnetic pole or south magnetic pole, but not both.

muon is a particle like the electron except that its mass is 207 times that of the electron and it is unstable.

neutrino is a massless particle that interacts weakly and was postulated by Pauli to explain energy conservation in beta decay.

nucleon is a constituent of the nucleus. it is either a proton or a neutron.

parsec is a unit of distance. It is the distance at which the average diameter of the earth's distance from the sun would subtend an angle of one second of arc. it corresponds to 3.26 light years.

parton a hypothetical point-like particle that is a constituent of hadrons. It is a generic quark.

pion is the particle that mediates the strong force between nucleons.

positron is the positively charged electron.

QED stands for quantum electrodynamics.

renormalization is a procedure in which the calculated values of certain quantities are replaced by their measured values so as to render all the quantities in the theory finite.

superconductivity is a phenomenon displayed by certain materials that their electrical reistivity vanishes below a certain critical temperature.

superconductor is any material that displays superconductivity.

superfluid is a fluid with no viscosity.

supersymmetry a theory of elementary particles in which the spin of a particle is used. The theory relates fermions with half-odd integral spin to boson with integral spin.

weak force is the force responsible for beta decay.

12.2   Chronology of Selected Physics Events

1912 Victor Hess shows that cosmic rays originate from outside the earth's atmosphere.

1915 Einstein publishes his Theory of General Relativity.

1916 Karl Schwarzschild produces an exact vacuum solution of the Einstein field equations for a spherically symmetric non-rotating system.

1928: Paul Adrien Maurice Dirac publishes his famous relativistic wave equation for the electron.

1928: George Gamow explains that alpha emission is caused by quantum tunneling.

1928: Walther Bothe and Hans Becker bombard beryllium with alpha particles and discover nuclear disintegration.

1929: Ernest Lawrence builds the first cyclotron.

1929: Paul Dirac postulates that all the negative energy states are filled — the so-called "Dirac sea" and propses hole theory.

1929: Edwin Hubble publishes his first estimate of the Hubble constant. His measurements implied that the universe is expanding.

1931: Paul Dirac, J Robert Oppenheimer, and Hermann Weyl predict the positron.

1931: Albert Einstein discards the cosmological constant.

1931: Wolfgang Pauli proposes the neutrino to explain energy conservation in beta decay.

1931: Paul Dirac proposes the existence of magnetic monopoles to explain the quantization of charge.

1932 Arthur Compton verifies the latitude effect and shows that cosmic rays consist of charged particles.

1932 James Chadwick identifies the neutron.

1932 Carl Anderson observes the positron in cosmic rays.

1932 Werner Heisenberg proposes that the proton and neutron are constituents of the atomic nucleus and introduces isospin.

1933 Patrick M. S. Blackett and Giuseppe B Occhialini observe electron-positron pair creation.

1933 Otto Stern measures the magnetic moment of the proton.

1933 Fritz Zwicky introduces the concept of dark cold matter.

1934 Enrico Fermi publishes the four fermion theory of weak interactions and beta decay.

1935 Subrahmanyan Chandrasekhar calculates the mass limit—the "Chandrasekhar limit"—for collapse of a white dwarf star.

1936 Bohr proposes the compound nucleus.

1936 Carl Anderson and Seth Neddermeyer discover the muon in cosmic rays.

1937 Pyotr Kapitza discovers superfluidity of helium II.

1938 Otto R Frisch and Lise Meitner develop the theory of nuclear fission.

1939 Rudolf Peierls derives the so-called "Peierls equation" for the critical mass for a self-sustaining chain reaction.

1939 Robert Oppenheimer and Hartland Snyder show that gravitational collapse of a pressureless homogeneous fluid sphere leads to formation of a black hole.

1941 Lev Davidovich Landau produces a theory of superfluids.

1942 Enrico Fermi produces the first self-sustaining nuclear fission reaction.

1945 The first atom bomb is exploded.

1946 Marcello Conversi, Ettore Pancini, and Oreste Piccioni demonstrate that the muon is not the mediator of the nuclear force.

1946 George Gamow proposes the Big Bang Theory.

1947 Cecil F. Powell and Giuseppe B Occhialini find the negative pion.

1947 Willis E. Lamb Jr. measures the fine structure of the hydrogen atom—the "Lamb shift".

1947 Hans Bethe calculates the Lamb shift.

1948 Sin-Itiro Tomonaga, Julian Schwinger, and Richard Feynman independently develop the renormalization theory for quantum electrodynamics.

1948 Ralph A. Alpher, Hans A. Bethe, and George Gamow explain nucleosynthesis in the big bang. They also predict the cosmic blackbody radiation.

1948 Richard Feynman publishes his path integral approach to quantum theory.

1948 John Bardeen, Walter H. Brattain, and William B. Shockley invent the transistor.

1948 Freeman Dyson demonstrates the equivalence of the Schwinger and Feynman versions of QED.

1949 Otto Haxel, Johannes H. D. Jensen, Maria Goeppert-Mayer, and Hans E. Suess propose the nuclear shell model.

1949 Fred Hoyle introduces the term "big bang".

1950 Donald A. Glaser invents the bubble chamber and Luis W. Alvarez uses the hydrogen bubble chamber.

1953 Murray Gell-mann and independently T. Nakano and Kazuhiko Nishijima introduce the strangeness quantum number.

1953 Charles Townes develops the maser.

1956 T. D. Lee and C. N. Yang suggest the possibility that parity is violated in weak interactions.

1957 Parity violation is found by Chien Shung Wu in beta decay.

1957 John Bardeen, Leon N. Cooper, and John R. Schrieffer publish the BCS theory of superconductivity.

1957 Richard Feynman and Murray Gell-Mann and independently Robert E. Marshak and E. C. George Sudarshan publish the $V - A$ theory of weak interactions.

1958 Charles Townes and Arthur Shawlow develop the theory of the laser.

1960 Pound and Rebka measure the gravitational red-shift.

1961 Murray Gell-mann and Yuval Ne'eman introduce the eight-fold way and quarks into hadron theory.

1962 Brian Josephson publishes the theory of the Josephson effect.

1964 Peter Higgs, Robert Brout, and Francois Englert introduce the mechanism of symmetry breaking—the"Higgs mechanism".

1964 Val L. Fitch and James W. Cronin find that CP is not conserved in weak interactions.

1964 John Bell publishes his inequalities that allow a test for local hidden variables.

1965 Arno A. Penzias and Robert W. Wilson detect the cosmic background radiation.

1965 Robert H. Dicke and Jim Peebles identify the cosmic background radiation detected by Penzias and Wilson.

1967 Steven Weinberg, Abdus Salam, and Sheldon Glashow independently achieve a unification of the electromagnetic and weak interaction to produce the electroweak theory.

1967 Werner Israel presents a proof of the "no hair" theorem.

1968 Joseph Weber makes a first (unsuccessful) attempt to measure gravitational waves.

1969 Richard Feynman produces the parton model.

1969 Jerome E. Friedman, Henry W. Kendall, and Richard E. Taylor find the parton structure inside protons in deep inelastic scattering.

1971 Cygnus X-1/HDE 226868 is identified as a possible binary black hole system.

1973 Julius Wess and Bruno Zumino develop the idea of supersymmetry for quantum field theory.

1974 Ting and Richter find the J/ψ particle.

1977 Klaus von Klitzing discovers the quantum Hall effect.

1982 Alain Aspect tests the Bell inequalities experimentally and confirms quantum mechanics and the non-existence of local hidden variables.

1983 Carlo Rubbia and co-workers at CERN detect the W and Z bosons.

1986 J. George Bednorz and K. Alexander Müller discover the first high temperature superconductor.

Index